The Genetics and Development of Scoliosis

Kenro Kusumi • Sally L. Dunwoodie
Editors

The Genetics and Development of Scoliosis

Second Edition

Editors
Kenro Kusumi
School of Life Sciences
Arizona State University
Tempe, AZ, USA

Sally L. Dunwoodie
Developmental and Stem Cell Biology Division
Victor Chang Cardiac Research Institute
Sydney, Australia

ISBN 978-3-030-07944-4 ISBN 978-3-319-90149-7 (eBook)
https://doi.org/10.1007/978-3-319-90149-7

Softcover re-print of the Hardcover 2nd edition 2018

Printed on acid-free paper

This Springer imprint is published by the registered company Springer International Publishing AG part of Springer Nature.
The registered company address is: Gewerbestrasse 11, 6330 Cham, Switzerland

The editors thank the patients and their families who have participated in scoliosis genetic studies and the research collaborators who have made these efforts possible

Preface

Scoliosis is a lateral curvature of the spine that is frequently encountered by healthcare professionals. Scoliosis has historically been categorized into congenital, neuromuscular, and idiopathic forms, and related curves include kyphosis, kyphoscoliosis, and lordosis. Patients affected by scoliosis are concerned about prognosis, associated health conditions, and recurrence risks. Developmental genetic studies of the spine and next-generation sequencing-based genetic analysis have led to recent advances in understanding the genetic etiology of idiopathic and congenital scoliosis.

The inspiration for the First Edition was derived from the invited session, Straightening Out the Curves: Understanding the Genetics Basis of Idiopathic and Congenital Scoliosis organized at the 2008 American College of Medical Genetics, Annual Clinical Genetics Meeting in Phoenix, AZ, USA. The Second Edition presents significant progress in understanding the genetic etiology of adolescent idiopathic scoliosis, presented March 16–17, 2017, at the Genomic Approaches to Understanding and Treating Scoliosis Conference, in Dallas, TX, USA. This meeting was a joint session of the International Consortium for Vertebral Anomalies and Scoliosis and the International Consortium for Scoliosis Genetics. These groups have now combined forces and merged into the International Consortium for Spinal Genetics, Development, and Disease.

Our understanding of the genetic and developmental mechanisms underlying idiopathic and congenital scoliosis is rapidly evolving, and our goal in editing *The Genetics and Development of Scoliosis*, Second Edition, was to provide researchers, clinicians, and students with the emerging views in this field.

Tempe, AZ, USA — Kenro Kusumi
Sydney, Australia — Sally L. Dunwoodie

Contents

Contributors

Peter G. Alexander Center for Cellular and Molecular Engineering, Department of Orthopaedic Surgery, University of Pittsburgh School of Medicine, Pittsburgh, PA, USA

Erin E. Baschal Department of Orthopedics, University of Colorado Anschutz Medical Campus, Aurora, CO, USA

Rebecca E. Fisher School of Life Sciences, Arizona State University, Tempe, AZ, USA

Department of Basic Medical Sciences, The University of Arizona College of Medicine– Phoenix, Phoenix, AZ, USA

Philip Giampietro Department of Pediatrics, Drexel University College of Medicine, Philadelphia, PA, USA

Ryan Scott Gray Department of Pediatrics, The University of Texas at Austin Dell Medical School, Austin, TX, USA

Shiro Ikegawa Laboratory for Bone and Joint Diseases, RIKEN Center for Integrative Medical Sciences, Tokyo, Japan

Zhaoyang Liu Department of Pediatrics, The University of Texas at Austin Dell Medical School, Austin, TX, USA

Ricardo Londono Center for Cellular and Molecular Engineering, Department of Orthopaedic Surgery, University of Pittsburgh School of Medicine, Pittsburgh, PA, USA

Thomas P. Lozito Center for Cellular and Molecular Engineering, Department of Orthopaedic Surgery, University of Pittsburgh School of Medicine, Pittsburgh, PA, USA

Jeremy McCallum-Loudeac Department of Anatomy, University of Otago, Dunedin, New Zealand

Nancy Hadley Miller Department of Orthopedics, University of Colorado Anschutz Medical Campus, Aurora, CO, USA

Musculoskeletal Research Center, Children's Hospital Colorado, Aurora, CO, USA

Alan Rawls School of Life Sciences, Arizona State University, Tempe, AZ, USA

Kazuki Takeda Laboratory of Bone and Joint Diseases, Center for Integrative Medical Sciences, RIKEN, Tokyo, Japan

Department of Orthopedic Surgery, Keio University School of Medicine, Tokyo, Japan

Elizabeth A. Terhune Department of Orthopedics, University of Colorado Anschutz Medical Campus, Aurora, CO, USA

Rocky S. Tuan Center for Cellular and Molecular Engineering, Department of Orthopaedic Surgery, University of Pittsburgh School of Medicine, Pittsburgh, PA, USA

Peter D. Turnpenny Clinical Genetics Department, Royal Devon & Exeter NHS Foundation Trust, Exeter, UK

University of Exeter Medical School, Exeter, UK

Megan J. Wilson Department of Anatomy, University of Otago, Dunedin, New Zealand

Carol A. Wise Sarah M. and Charles E. Seay Center for Musculoskeletal Research, Texas Scottish Rite Hospital for Children, Dallas, TX, USA

Departments of Orthopaedic Surgery, Pediatrics, and McDermott Center for Human Growth and Development, University of Texas Southwestern Medical Center, Dallas, TX, USA

Nan Wu Department of Orthopedic Surgery, Peking Union Medical College Hospital, Peking Union Medical College and Chinese Academy of Medical Sciences, Beijing, China

Beijing Key Laboratory for Genetic Research of Skeletal Deformity, Beijing, China

Medical Research Center of Orthopedics, Chinese Academy of Medical Sciences, Beijing, China

Chapter 1
Developmental and Functional Anatomy of the Spine

Alan Rawls and Rebecca E. Fisher

Introduction

The vertebral column is composed of alternating vertebrae and intervertebral (IV) discs supported by robust spinal ligaments and muscles. All of these elements, bony, cartilaginous, ligamentous, and muscular, are essential to the structural integrity of the spine. The spine serves three vital functions: protecting the spinal cord and spinal nerves, transmitting the weight of the body, and providing a flexible axis for movements of the head and torso. The vertebral column is capable of extension, flexion, lateral (side to side) flexion, and rotation. However, the degree to which the spine is capable of these movements varies by region. These regions, including the cervical, thoracic, lumbar, and sacrococcygeal spine, form four curvatures (Fig. 1.1). The thoracic and sacrococcygeal curvatures are established during the fetal period while the cervical and thoracic curvatures develop during infancy. The cervical curvature is established in response to holding the head upright, while the lumbar curvature develops as an infant begins to sit upright and walk. However, congenital defects and degenerative diseases can result in exaggerated, abnormal curvatures. The most common of these include kyphosis (hunchback deformity), lordosis (swayback deformity), and scoliosis. Scoliosis involves a lateral curvature of greater than 10 °, often accompanied by a rotational defect. To appreciate the potential underlying causes of scoliosis, we need to understand the cellular and genetic basis of spinal development and patterning. In this chapter, we will review the embryonic

---

A. Rawls
School of Life Sciences, Arizona State University, Tempe, AZ, USA

R. E. Fisher (✉)
School of Life Sciences, Arizona State University, Tempe, AZ, USA

Department of Basic Medical Sciences, The University of Arizona College of Medicine–Phoenix, Phoenix, AZ, USA
e-mail: rfisher@email.arizona.edu

K. Kusumi, S. L. Dunwoodie (eds.), *The Genetics and Development of Scoliosis*, https://doi.org/10.1007/978-3-319-90149-7_1

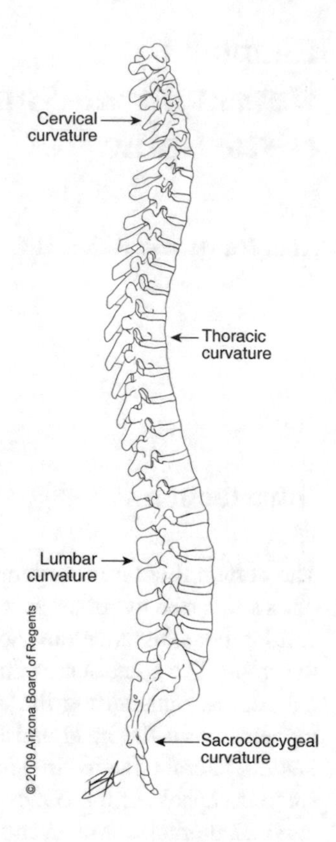

Fig. 1.1 Lateral view of the vertebral column, illustrating the spinal curvatures (Drawing by Brent Adrian)

development of the spine and associated muscles and the functional anatomy of these structures in the adult.

Embryonic Origins of the Spine

The origins of the vertebral column, spinal musculature, and associated tendons are two rods of paraxial mesoderm that fill in the space on either side of the neural tube at the time of gastrulation. Beginning at 20 days *post coitus*, paraxial mesoderm undergoes segmentation in a rostral to caudal direction to form 42–44 pairs of somites, which can be subdivided into 4 occipital, 8 cervical, 12 thoracic, 5 lumbar, 5 sacral, and 8–10 coccygeal somites. The first occipital and the last 5–7 coccygeal somites disappear during embryonic development. Each somite will differentiate into four cell lineage-specific compartments that contribute to the vertebral column and associated musculature, including the sclerotome (vertebrae and ribs), syndetome (tendons), myotome (skeletal muscle), and dermomyotome (dermis and skeletal muscle progenitor cells).

Somite formation can best be described as a continuous segmentation of mesenchymal cells from the rostral end of the paraxial mesoderm or presomitic mesoderm (PSM) that lays down the embryonic cells that will give rise to the axial skeleton. Intrinsic to this process is (1) an oscillating clock controlling the timing of somitogenesis, (2) the formation of intersomitic boundaries, (3) mesenchymal to epithelial transition (MET), and (4) positional identity (e.g., rostral/caudal and dorsal/ventral). Experimental disruption in any one of the processes in vertebrate model organisms (e.g., mouse and chick) can lead to an axial skeletal dysmorphogenesis that is phenotypically consistent with scoliosis. The timing of somite formation and the determination of the site of boundary formation are established by the interactions between the Notch, Wnt, and FGF signaling pathways. Here we will focus on the morphogenetic events associated with the physical separation of PSM during formation of the boundary, epithelialization, and positional identity.

Establishing the Intersomitic Boundary

Boundary formation occurs as somitic cells pull apart from the adjacent PSM. Dependent on the animal, this varies from the simple cleavage of the PSM by fissures initiated along either the medial or lateral surfaces as seen in *Xenopus* and zebra fish to a more dynamic ball-and-socket shape with a reshuffling of cells across the presumptive somite-PSM boundary in chicks [57, 64, 74, 75, 163]. The activity is an intrinsic property of the PSM, as it will occur in explants in the absence of the adjacent ectoderm and endoderm [108]. However, the underlying mechanism(s) remains poorly understood. In studies carried out in chick embryos, the fissure can be induced by activated Notch receptors and is stabilized by the presence of Lfng [128]. Transcription factors *Mesp2* (and its chicken homologue, *cMeso1*) and *Tbx18* have also been shown to play a role in forming boundaries [19, 124, 146, 152]. Ectopic expression of either *cMeso1* or *Tbx18* is sufficient to induce ectopic fissures in chick PSM. Additional signals derived from the ventral PSM coordinate fissure formation in the dorsal PSM, though the nature of the signal remains poorly understood [127]. It is likely that the physical separation of cells at the fissure is related to differential changes in cell adhesion.

Somite Epithelialization

Cells of the newly formed somites undergo an increase in cell number, density, and expression of extracellular matrix proteins (reviewed in [70, 151]), resulting in the condensation of mesenchyme into an epithelial ball, surrounding a mesenchymal core, called the somitocoele. This occurs in a gradual process with the cells along the rostral edge of somite 0 becoming epithelia at the time of boundary formation [46]. Epithelialization is complete with the formation of the next boundary (Fig. 1.2).

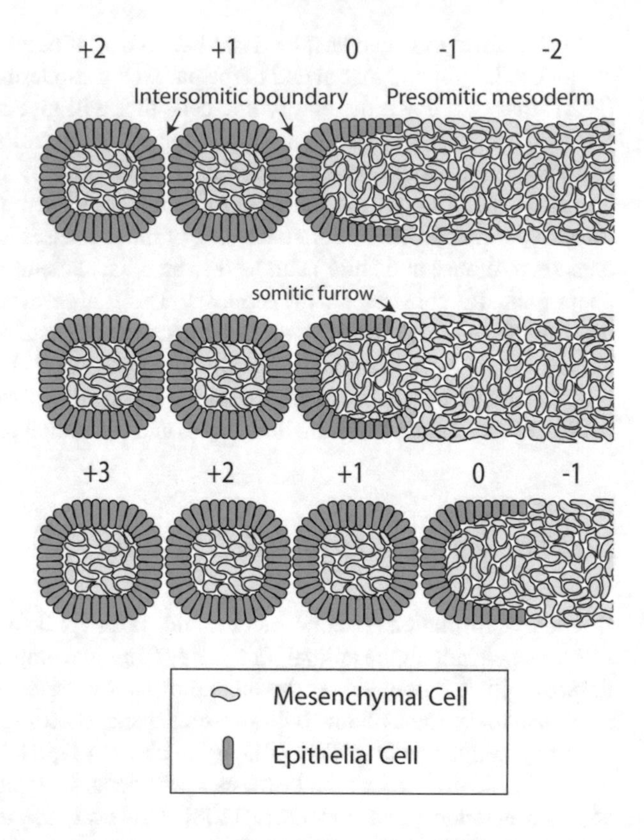

Fig. 1.2 Schematic of mouse somite formation. Lateral view of somites budding off the rostral end of the presomitic mesoderm demonstrates the stepwise transition of mesenchymal cells to epithelium. By convention, the forming somite is labeled "0" and the newest somite is "+1"

The transcription factors paraxis and *Pax3* are required to direct MET in cells of somite +1 [20, 21, 86, 130]. Inactivation of paraxis results in somites formed of loose clusters of mesenchyme separated by distinct intersomitic boundary formation (Fig. 1.2). This reveals that MET is not required for boundary formation. However, the two events are temporally linked, suggesting that they are both responsive to the oscillating segmental clock. Candidate genes for linking the two are *snail1* and *snail2* (*Snai1* and *Snai2*), which are expressed in oscillating patterns in the PSM [40]. Snail genes are transcriptional repressors that are able to block the transcription of paraxis and cell adhesion molecules associated with epithelialization [9, 10, 26, 40]. Overexpression of *Snai2* will prevent cells from contributing to epithelium in somite +1. Thus, switching off snail gene expression may be essential for the timing of MET.

In contrast to boundary formation, signals from the surface ectoderm are required to induce MET and the expression of paraxis [38, 45, 80, 127, 128, 138]. Wnt signaling has been implicated in regulating this process with *Wnt6* and *Wnt11* as the most likely candidates [55, 80, 129, 159]. Ectopic expression of *Wnt6* is able to rescue somite epithelialization where the ectoderm has been removed. Further, *Wnt6* is able to induce paraxis transcription through a beta-catenin-dependent manner, predicting a mechanism of action [80].

Somite epithelialization is associated with an increase in the expression of members of the cadherin superfamily and cell adhesion molecules [45, 151]. These cell surface molecules participate in the formation of focal adhesion and desmosomes at the apical junction of epithelium. Inactivation of N-cadherin (*Cdh2*), alone or in combination with cadherin 11 (*Cdh11*), leads to the disorganization of the somite epithelium into small clusters of cells [58, 79, 116]. Functional inactivation of *Cdh2* through increased endocytosis has been implicated in the formation of the new somitic boundary. The protocadherin, PAPC, which is dynamically expressed in the forming somites regulated by Notch/Mesp2 signaling, promotes clathrin-mediated endocytosis and the internalization of Cdh2 [29, 119]. This disrupts homotypic interaction of cadherins between adjacent cells leading to a fissure that will become the somitic boundary.

The phenotypes of the cadherin mutations are not as severe as either the paraxis or *Pax3*, predicting that additional factors associated with cell adhesion are required for epithelialization. The most likely candidates are the genes involved in cytoskeletal remodeling. Likely targets are members of the Rho family of GTPase. In the chick, overexpression of *Cdc42* promotes somitic cells to maintain their mesenchymal state [103]. Both the inhibition and over-activation of Rac1 disrupt somite epithelialization, demonstrating the sensitivity of the cells to disruption of this pathway. The activity of Rac1 cannot be rescued by paraxis predicting that Rac1 is acting downstream [103]. In the paraxis-null, localization of Rac1 is disrupted in the somites, and the regulation of the expression of Rac1 modifiers, including the guanine nucleotide exchange factor, Dock2, is disrupted reinforcing a role for paraxis downstream of Rac1 [123].

Differential gene expression studies with paraxis-null somites revealed a significant reduction in the expression of fibroblast activation protein alpha (*Fap*), encoding a dipeptidyl peptidase that regulates fibronectin and collagen fiber organization in extracellular matrix [123]. Further, downstream genes in the Wnt and Notch signaling pathways were downregulated in the absence of paraxis, predicting a positive feedback loops with both pathways.

Rostral/Caudal Polarity of Somites

Spatial identity along the rostral/caudal axis is established in each somite at the time of its formation [3, 56]. Rostral/caudal polarity is essential for imposing the segmental patterning of the peripheral nerves and the resegmentation of the sclerotome during vertebrae formation. This is regulated by an intricate feedback loop between cells in the rostral and caudal halves of the forming somite (somite 0). Consistent with the cyclical nature of somitogenesis, the feedback loop is also entrained with the oscillating segmental clock. Activation of the Notch pathway plays a central role in determining spatial identity. Disruption of *Notch1*, ligands *Dll1* and *Dll3*, or modifying gene peptide-O-fucosyltransferase 1 (*Pofut1*) and presenilin-1 lead to the loss of rostral- and caudal-specific gene expression, fusion of the vertebrae, and disruption of the segmental pattern of the peripheral nerves [41, 47, 59, 73, 76, 104, 131, 144]. Spatial identity of the rostral half of the somite requires the expression of *Mesp2*, which is transcribed in a broad domain that encompasses presumptive somite −1 before becoming restricted to the rostral half of the presumptive somite (somite 0) [124, 147]. Mouse embryos deficient in *Mesp2* lead to expanded expression of caudal-specific genes and fused vertebrae. Transcription of *Mesp2* is upregulated by activated Notch in a *Tbx6*-dependent manner [166], which in turn represses transcription of the *Dll1* ligand in the rostral domain through the transcriptional repressor, Ripply2 [101]. In the caudal half of somite 0, *Mesp2* transcription is repressed by a presenilin-1-dependent manner [73, 148, 166].

Maintenance of rostral/caudal polarity after somite formation requires paraxis, which is associated with the regulation of somite epithelialization [65]. In paraxis-null embryos, the transcription pattern of *Mesp2* and components of the Notch signaling pathway are unaltered in somite 0 and − 1. However, the expression of caudal-specific genes, such as *Dll1* and *Uncx4.1*, is broadly transcribed in the newly formed somites. It has been proposed that paraxis participates in a cell adhesion-dependent mechanism of maintaining the intersomitic boundary between the rostral and caudal halves of the somite after their specification in the presomitic mesoderm [65].

The Anatomy and Development of the Vertebrae and Intervertebral Discs

A typical vertebra consists of two parts: the body and the vertebral (or neural) arch (Fig. 1.3A). The vertebral body is located anteriorly and articulates with the adjacent intervertebral (IV) discs (Figs. 1.1, 1.3, and 1.4). Together, the vertebral body and arch form a central, vertebral foramen, and, collectively, these foramina create a vertebral canal that protects the spinal cord. In this section, the functional anatomy of the vertebrae and IV discs in the adult and the genetic basis for their development in the embryo will be discussed.

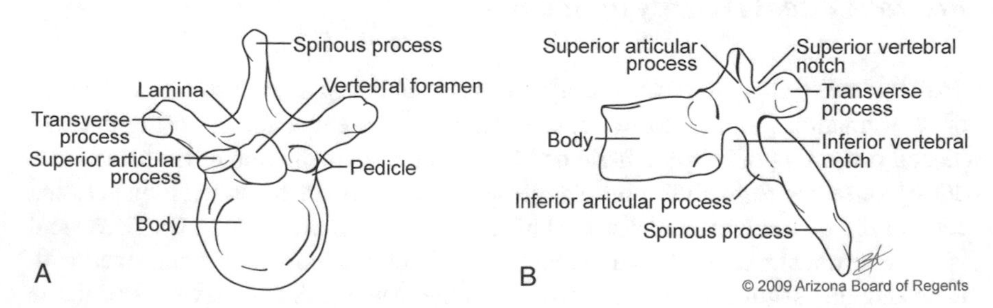

Fig. 1.3 Features of a typical human vertebra. (**A**) Superior and (**B**) lateral view (Drawing by Brent Adrian)

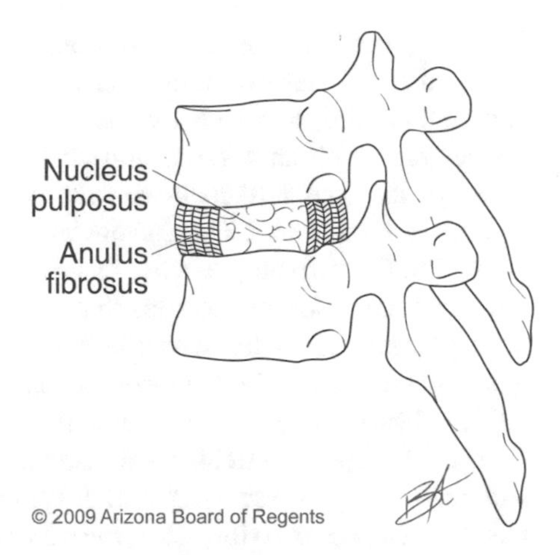

Fig. 1.4 Structure of the intervertebral disc (Drawing by Brent Adrian)

Functional Anatomy of the Vertebrae and IV Discs

The vertebral bodies consist of a shell of compact bone surrounding a core of trabecular bone and red marrow. In addition, hyaline cartilage forms vertebral end plates on the superior and inferior surfaces of each body. The vertebral bodies, in conjunction with the IV discs, bear and transmit weight; as a result, the bodies increase in size from the cervical to the lumbar region (Fig. 1.1). However, as weight is then transferred to the lower extremities via the sacrum, the bodies subsequently decrease in size.

The vertebral arch is located posterior to the vertebral body and consists of two pedicles and two laminae (Fig. 1.3A). The superior and inferior notches of adjacent pedicles form the intervertebral foramina, which transmit the spinal nerves (Figs. 1.1 and 1.3B). Disruption of these foramina (e.g., by a herniated disc) can compress the spinal nerves, leading to both sensory and motor deficits. In addition to protecting the spinal cord and spinal nerves, the vertebral arch also has a number of processes that provide sites for muscle and ligament attachment. The spinous process, located at the junction of the laminae, and the transverse processes, located at the pedicle-lamina junctions, provide attachment sites for ligaments as well as the erector spinae and transversospinalis muscles (Fig. 1.3A–B). In addition, in the thoracic region, the transverse processes articulate with the tubercles of the ribs to form the costovertebral joints. Finally, the superior and inferior articular processes of adjacent vertebrae interlock to form the zygapophysial (or facet) joints (Fig. 1.4). These synovial joints permit gliding movements and their orientation largely determines the ranges of motion that are possible between adjacent vertebrae.

The morphology and the functions of the vertebrae vary by region. The cervical spine is composed of seven vertebrae (Fig. 1.1). The bodies are small, reflecting their relatively minor weight-bearing role, while transverse foramina are present for the passage of the vertebral arteries and veins. In addition, the articular facets on the superior and inferior articular processes face superiorly and inferiorly, promoting flexion, extension, lateral flexion, and rotation at the cervical facet joints. This region also includes two highly derived elements, the C1 and C2 vertebrae. The C1 vertebra, or atlas, lacks a body and spinous process. Instead, it features two lateral masses united by an anterior and posterior vertebral arch. The superior articular facets of the atlas articulate with the occipital condyles of the skull to form the atlanto-occipital joints. These synovial joints allow for flexion and extension of the head. The C2 vertebra, or axis, features a dens or odontoid process; this process represents the body of the atlas that fuses with the axis during development. The dens process articulates with the anterior arch of the atlas to form the median atlanto-axial joint while the facet joints between the C1 and C2 vertebrae form the lateral atlanto-axial joints. Together, these joints allow for rotation of the head.

The 12 thoracic vertebrae are distinct in featuring costal facets on their bodies and transverse processes (Fig. 1.3B). Typically, a thoracic vertebral body articulates with the heads of two ribs, while the transverse process articulates with the tubercle of one of these ribs; altogether, these articulations form the costovertebral joints. These synovial joints serve to elevate and depress the ribs, thus increasing the anterior-posterior and transverse diameters of the thoracic cavity during respiration. In the thoracic spine, the superior and inferior articular facets face anteriorly and posteriorly (Fig. 1.3B), permitting rotation and some lateral flexion. However, the orientation of these facets, as well as the inferiorly directed spinous processes and the costovertebral joints, severely restricts flexion and extension of the thoracic spine. In contrast, the medially and laterally facing articular facets of the five lumbar vertebrae allow for a great deal of flexion and extension, but restrict rotation. The lumbar vertebrae also exhibit robust vertebral bodies and well-developed spinous, transverse, and articular processes that provide attachment sites for ligaments as well as the erector spinae and transversospinalis muscles (Fig. 1.1).

The sacrum is typically formed by the fusion of five sacral vertebrae (Fig. 1.1). The sacral canal transmits the spinal roots of the caudal equina and ends at the sacral hiatus, an important landmark for administering a caudal epidural. In addition, pairs of sacral foramina transmit the ventral and dorsal rami of the sacral spinal nerves. The sacrum plays an important role in transmitting the weight of the body from the spine to the lower extremities; as a result, the sacroiliac joints are protected by extremely robust ligaments. The coccyx is typically formed by the fusion of four coccygeal vertebrae (Fig. 1.1). Although rudimentary in humans, the coccyx serves as a focal point for the attachment of the muscles of the pelvic floor as well as the sacrotuberous and sacrospinous ligaments.

Most of the vertebral bodies articulate superiorly and inferiorly with IV discs, forming secondary cartilaginous joints or symphyses (Fig. 1.4). However, an IV disc is not present between the atlas and axis, and the sacral and coccygeal IV discs ossify progressively into adulthood. Representing up to 25% of the total length of the spine, the IV discs act as shock absorbers and enhance spinal flexibility, particularly in the cervical and lumbar regions [100]. The IV discs are responsible for resisting compressive loads due to weight bearing as well as tensile and shearing stresses that arise with movements of the vertebral column, such as rotation and lateral flexion. The thoracic IV discs are relatively thin and uniform in shape, while

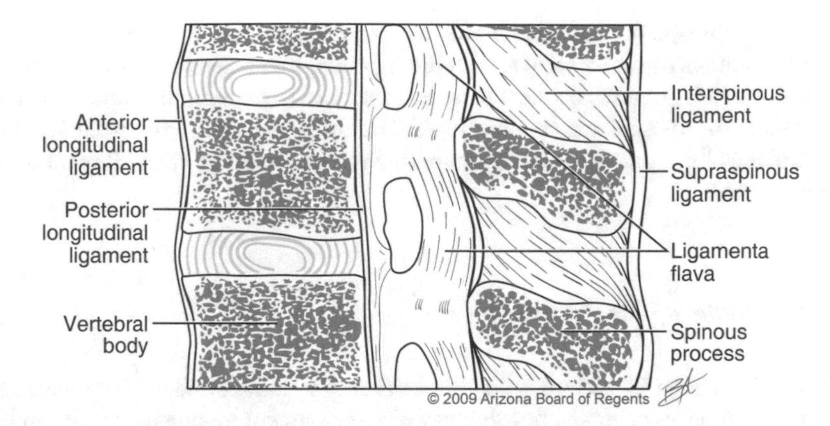

Fig. 1.5 Major ligaments of the spine. Lateral view illustrating the ligamentum flava, supraspinous, interspinous, and anterior and posterior longitudinal ligaments (Drawing by Brent Adrian)

the cervical and lumbar IV discs are wedge-shaped, contributing to the curvatures of the vertebral column (Fig. 1.1). Each IV disc is composed of an outer fibrocartilaginous ring, the annulus fibrosus, and a central gelatinous core, the nucleus pulposus (Fig. 1.4). Composed primarily of collagen fibers, the annulus fibrosus is characterized by a series of concentric layers, or lamellae (Fig. 1.4). The lamellae serve to resist the expansion of the nucleus pulposus during compression [25]. The nucleus pulposus is composed of water, proteoglycans, and scattered collagen fibers.

The vertebrae and IV discs are stabilized by robust spinal ligaments that function to restrict movements and to minimize the need for continual muscular contraction. The major spinal ligaments are illustrated in Fig. 1.5. The broad anterior longitudinal ligament is situated on the anterior surface of the vertebral bodies and IV discs and extends from the sacrum to the occipital bone (Fig. 1.5). This ligament, which prevents hyperextension of the spine and anterior herniation of the nucleus pulposus, is especially prone to injury in the cervical region due to whiplash (hyperextension) injuries. The posterior longitudinal ligament is slender compared to its counterpart. It lies within the vertebral canal, on the posterior surface of the vertebral bodies and IV discs (Fig. 1.5). This ligament prevents hyperflexion of the vertebral column and posterior herniation of the nucleus pulposus. In fact, due to the presence of the posterior longitudinal ligament, the nucleus pulposus tends to herniate in a posterolateral direction.

While the anterior and posterior longitudinal ligaments traverse the length of the spine, the ligamenta flava connect the laminae of adjacent vertebrae (Fig. 1.5). These ligaments contribute to the posterior wall of the vertebral canal, thus helping to protect the spinal cord. The ligamenta flava are highly elastic, supporting the normal curvatures of the spine, resisting separation of the laminae during flexion, and assisting in extending the spine from a flexed position. The vertebrae are also held together by the intertransverse and interspinous ligaments, which connect adjacent transverse and spinous processes, respectively (Fig. 1.5). More superficially, the robust supraspinous ligament binds the spinous processes together. In the neck, the supraspinous ligament merges with the ligamentum nuchae, a fibroelastic structure that extends from the cervical spinous processes to the occiput, forming a midline raphe for muscle attachment [94]. The intertransverse, interspinous, and supraspinous ligaments help prevent hyperflexion and extreme lateral flexion of the vertebral column.

Development of the Vertebrae

The axial skeleton is derived from the sclerotome compartment of the somites, which first appear during the fourth week of development in humans as the epithelial cells in the ventral/medial quadrant of the somite undergo an epithelial-to-mesenchymal transition (EMT). These cells, along with the mesenchymal cells of the somitocoele, are initially specified to the chondrogenic lineage and form the cartilage models of the vertebrae (Fig. 1.6) (reviewed in [44]). Through endochondral ossification, the cartilage is replaced by bone. The molecular events that regulate this process are similar to those that regulate the appendicular skeleton and part of the cranium. These pathways are reviewed elsewhere [81]. In this chapter, we will focus on the signaling events that influence patterning of the newly formed vertebrae.

The transition from sclerotome to vertebrae can be divided into distinct processes for the ventral structures (vertebral body and intervertebral discs) and dorsal neural arch structures (pedicles, laminae, spinous and transverse processes) based on both cell origin and genetic regulation. Patterning along the dorsal/ventral axis is controlled by opposing gradients derived from the notochord and surface ectoderm overlying the neural tube. Sonic hedgehog (*Shh*) and the BMP inhibitor, noggin, have been identified as factors expressed in the notochord that are sufficient to promote the expression of the transcription factors *Pax1*, *Pax9*, and *Mfh1* in the sclerotome [49, 53, 91, 110]. *Pax1* and *Pax9* are essential for the maintenance of sclerotomal cells [53]. Compound mutations of these two genes in the mouse lead to loss of the vertebral body and proximal ribs [111]. In addition to signals from the notochord, the polycomb genes *Pbx1* and *Pbx2* and bHLH genes paraxis and *Mesp2* are also required for *Pax1* and *Pax9* transcription [28, 149]. The homeodomain-containing genes, *Meox1* and *Meox2*, that are essential for vertebrate development [84, 135] combine with *Pax1* and *Pax9* to activate the expression of *Nkx3.2*, a transcriptional repressor that triggers chondrogenesis [121, 122]. Chondrocyte differentiation is associated with the downregulation of *Pax1* in the cells of the sclerotome. Though *Pax1* is required for sclerotome specification, it is an inhibitor of chondrogenesis through the inhibition of *Sox9*, *Nkx3.2*, Indian hedgehog, and aggrecan [150]. This dual role for *Pax1* likely allows for further subdivision of the sclerotome as the cells that contribute the intervertebral disc and fail to undergo chondrogenesis maintain its expression.

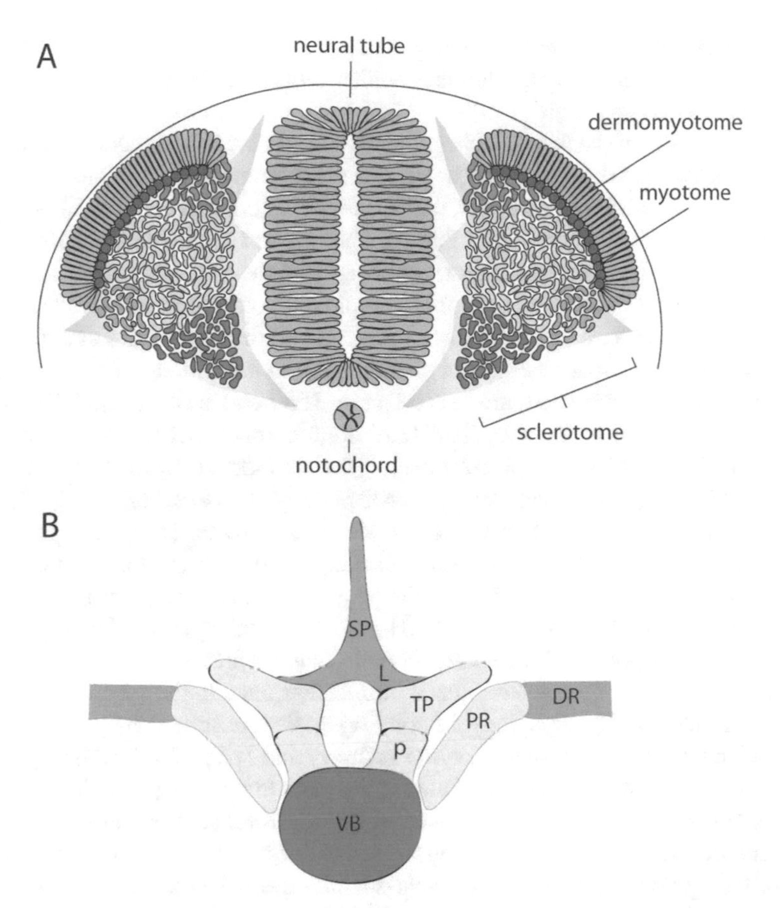

Fig. 1.6 Sclerotome origins of the vertebrae. (**A**) A schematic of the differentiating somites demarcating the domains of the sclerotome that migrate to form the individual elements of the vertebrae. (**B**) A diagram of a thoracic vertebra. Vertebral body (VB; green), pedicle (P; yellow), transverse process (TP; yellow), lamina (L; blue) and spinous processes (SP; blue), proximal rib (PR; yellow), and distal rib (DR; orange)

The formation of the vertebral body is dependent on the highly coordinated migration of sclerotomal cells both toward the midline and along the rostral/caudal axis (Fig. 1.6) (reviewed in [15]). Soon after EMT, cells from the ventral/medial sclerotome migrate toward the notochord. This is directed in part through an interaction with an extracellular matrix network (e.g., laminin, fibronectin, collagen I, aggrecan, and perlecan) radiating from the notochord [62]. Production of the matrix genes requires the expression of *Sox5* and *Sox6* [137]. Initially, the *Pax1 + ve* sclerotomal cells form an unsegmented sheath around the notochord that will give rise to both the future vertebral bodies and intervertebral discs. Segmentation appears as cells of the future intervertebral discs condense, and the intervening loose mesenchyme will give rise to the vertebral bodies [36]. The metameric pattern is also reflected in *Pax1* expression, which is maintained in the future intervertebral disc and lost from the vertebral body anlagen. This is believed to promote differential chondrocyte maturation in the vertebral bodies, while maintaining the intervertebral cell in a mesenchymal state [150].

The formation of the neural arches is more complicated as the pedicles and transverse processes are derived from the central sclerotome while the lamina and spinous processes originate from the dorsal/medial sclerotome (Fig. 1.6). They are further distinguished by their contribution from the rostral and caudal halves of the sclerotome, which are morphologically distinguishable at this time (Fig. 1.7A). The pedicles and transverse processes originate almost solely from the caudal domain and the spinous process from the rostral domain. While the pedicles and transverse processes are dependent on *Pax1* for specification to the chondrogenic lineage, the lamina and spinous processes are dependent on *Msx1* and *Msx2* transcription. Thus, these structures still develop in *Pax1/Pax9* double knockouts where the vertebrae are absent [111]. *Msx1* and *Msx2* transcription is induced by BMP2 and BMP4 expressed in the surface ectoderm and roof plate of the neural tube [98, 99, 160]. SHH and the BMP's are mutually antagonistic in their actions [113]. Ectopic expression of BMP2 or BMP4 on the dorsal neural tube will increase dorsal chondrogenesis while ectopic expression lateral to the neural tube inhibits chondrogenesis [154, 160]. The corollary is also true with SHH-expressing cells grafted dorsally, inhibiting *Msx1* transcription and preventing chondrogenesis [160].

Resegmentation of the sclerotome is intimately linked to the specification of the rostral and caudal domains early in somitogenesis. As described previously, the interaction between the Notch signaling pathway and *Mesp2* leads to the specification of the rostral and caudal fate of the somite prior to overt segmentation. As such, the caudalization of the somite by inactivation of *Mesp2* leads to fusion of the vertebral bodies and neural arches along the length of the vertebral column [124]. In contrast, disruption of the somites' caudal identity through inactivation of the Notch pathway leads to fused vertebral bodies and an absence of neural arches. Mutations in genes regulating this process have been identified as the cause of spondylocostal dysostoses, a heterogeneous group of disorders with severe axial skeletal malformation characterized radiographically by multiple vertebral segmentation defects (reviewed in [139]). Disruption of rostral/caudal polarity after somite formation has also been shown to impact resegmentation, though to a lesser extent. In *paraxis*-deficient embryos, ventral cartilage fails to segment into vertebral bodies and IV discs, while the lateral neural arches are unaffected [65].

Rostral/Caudal Patterning

An additional layer of regulation is required to confer the distinctive regional characteristics of the cervical, thoracic, lumbar, sacral, and caudal vertebrae. Members of the Hox transcription factor family have been strongly implicated in establishing positional identity of vertebrae along the rostral/caudal axis (reviewed in [161]). From classic studies in *Drosophila*, the Hox genes have long been known to

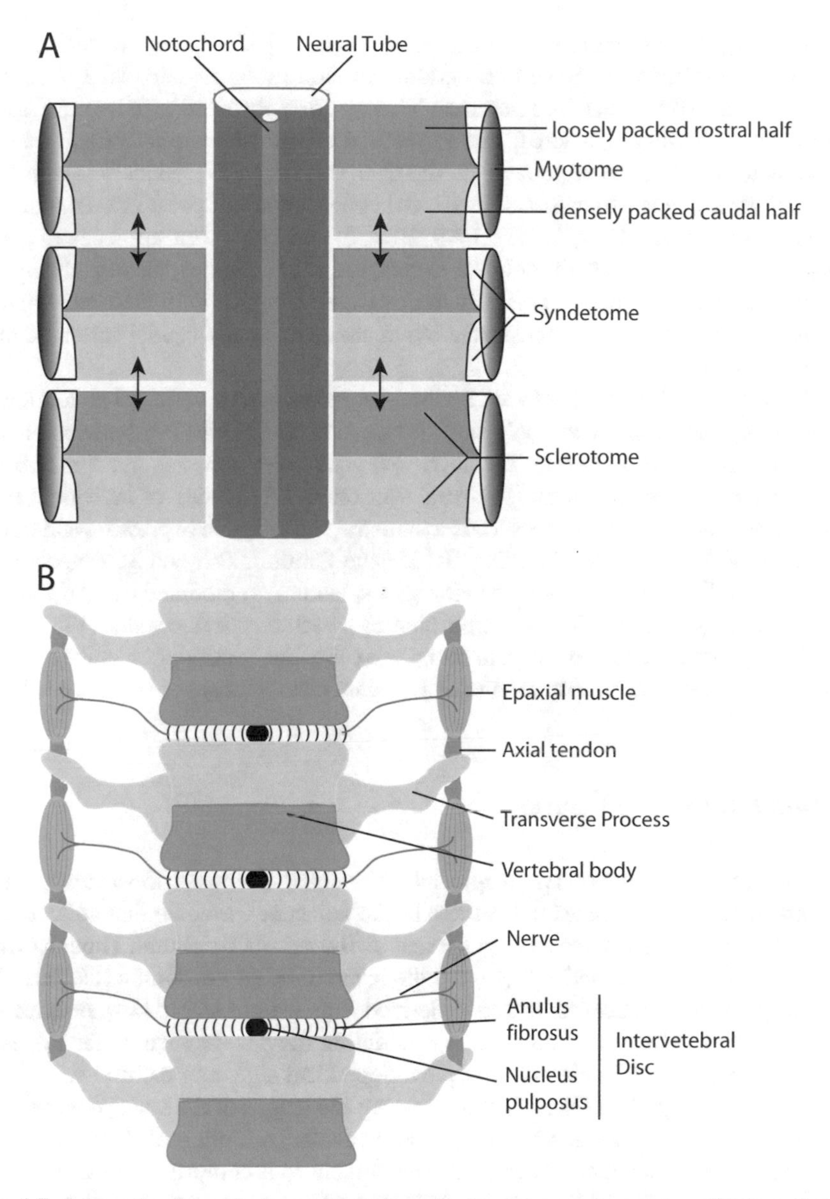

Fig. 1.7 Schematic of vertebral generation through sclerotome resegmentation. (**A**) Ventral view of the sclerotome, syndetome, and myotome compartments. The caudal half of the sclerotome grows into the rostral half of the adjacent somite. (**B**) Ventral view of the vertebral column with associated epaxial muscles and axial tendons. Shading represents the contribution of the rostral and caudal sclerotome to the vertebral bodies and transverse processes. The intervertebral disc forms at the site sclerotome separation. Note the relationship of the muscle and bone after resegmentation

regulate segmental identity in the insect body plan [77]. Compound mutations that inactivate more than one gene of a paralogous Hox group in mice lead to rostral homeotic transformation of the vertebrae. This was first observed with *Hoxa3/Hoxd3* double mutant embryos, where the prevertebral elements that normally contribute to the atlas form a bone contiguous with the occipital bone [37]. Since this observation, similar homeotic transformations have been reported for paralogous mutations in the Hox5, Hox6, Hox7, Hox8, Hox9, Hox10, and Hox11 group genes [31, 88, 156, 162]. Consistent with the colinear expression of these genes, the rostral homeotic transformations effect successively more caudal vertebrae with the Hox11 paralogous mutants displaying a transformation of sacral and early caudal vertebrae into a lumbar-like fate [162].

The positional identity conferred by the Hox genes during vertebral patterning is modified by members of the polycomb family and TALE class of homeodomain-containing transcription factors. The polycomb genes, *Bmi* and *Eed*, function as transcriptional repressors that limit the rostral transcription boundary of individual Hox genes. Inactivation of these genes leads to a rostral shift in gene expression and transformation of the vertebrae [72]. The TALE gene families, *Pbx* and *Meis* genes, are able to form dimer partners with the Hox genes, leading to modified transcription of target genes by altering DNA-specific binding specificity (reviewed in [96]). The TALE genes play a larger role in patterning, regulating the transcription of the 5 prime Hox genes by both a Hox-dependent and Hox-independent manner [11, 27, 82, 112].

Formation of the IV Discs

An IV disc is comprised of a proteoglycan-rich nucleus pulposus, the annulus fibrosus, and cartilage end plates that adhere to the adjacent vertebrae that collectively redistribute the compressive force generated by the vertebral column. Though originally thought to be derived solely from the sclerotome of the somite [60, 95], the nucleus pulposus has been shown to be derived from the notochord [34]. As a result, we must now invoke a more complicated model for the development of the IV discs that requires the coordination of multiple independent signaling pathways.

The notochord is a rodlike structure running the length of the embryonic ventral midline, where it serves as a signaling center for the patterning of the central nervous system, gut, and vertebral column. The notochord is comprised of highly vacuolated cells encapsulated in a sheath composed of collagen, aggrecan, fibronectin, laminin, cytokeratin, and sulfate glycosaminoglycans (GAGs). Components of the sheath including aggrecan and more than 100 sulfated GAGs are also found in the nucleus pulposus, where they maintain the osmolality essential for giving the tissue its gel-like characteristics [107, 134]. The signaling pathways that are required for nucleus pulposus formation remain poorly understood. Some insight has come from the study of *Shh*, which is required for the integrity of the notochordal sheath and cell proliferation [33]. In complete and conditional mutations, notochordal cells fail to properly migrate to the nucleus pulposus [35]. Sheath stability and ultimately maintenance of the notochord are dependent on *Sox5/Sox6* and *Foxa1/Foxa2* expression [83, 137]. Single mutations in either of the *Sox* or *Foxa* genes did not have notochord defects, suggesting functional redundancy of sister genes. In the case of *Foxa*, the proteins have been shown to bind to the *Shh* promoter predicting a role in regulating the Shh pathway [63].

The annulus fibrosus of an IV disc forms from condensed mesenchyme derived from the somitocoele at the border of the rostral and caudal domains during resegmentation [60, 95]. Somitocoele cells cannot be replaced by sclerotomal cells derived from EMT in forming the IV disc predicting specification of a distinct lineage, now called the arthrotome [95]. Development of the annulus fibrosus and its maintenance in adults is dependent on members of the TGF-beta superfamily. Inactivation of *TGF-beta type II receptor* (*Tgfbr2*) in type II collagen expressing cells results in an expansion of *Pax1/Pax9* expression and the loss of IV discs [8]. GDF-5 and BMP-2 promote cell aggregation and expression of the chondrogenic genes instead of osteogenic genes in the IV discs [78, 167].

The Anatomy and Development of Spinal Muscles

A number of muscle groups act upon the spine. Those located anterior to the vertebral bodies act as flexors, including longus capitis and colli, sternocleidomastoid, psoas major, and rectus abdominis. In contrast, the extensors of the spine are located posterior to the vertebral bodies and include the splenius, erector spinae, and transversospinalis muscles (Fig. 1.8). Lateral (side to side) flexion is achieved by the scalenes, sternocleidomastoid, splenius capitis and cervicis, and the erector spinae in the cervical region and quadratus lumborum, transversus abdominis, the abdominal obliques, and the erector spinae in the lumbar region. While flexors of the spine are innervated by the ventral rami of spinal nerves or the spinal accessory nerve (CN XI), the extensors are innervated by the dorsal rami of spinal nerves. Since the lateral flexors include members from both of these groups, their innervation varies. The term “spinal muscles” typically refers to the dorsal rami innervated splenius, erector spinae, and transversospinalis muscles. In this section, the functional anatomy of the spinal muscles and the genetic basis for their development in the embryo will be discussed.

Functional Anatomy of the Spinal Muscles

Splenius capitis and cervicis occupy the posterior aspect of the cervical region, deep to the trapezius and the rhomboids (Fig. 1.8A). They take origin from the ligamentum nuchae and cervical and thoracic spinous processes and insert onto the mastoid process and occipital bone (capitis) or the cervical transverse processes (cervicis)

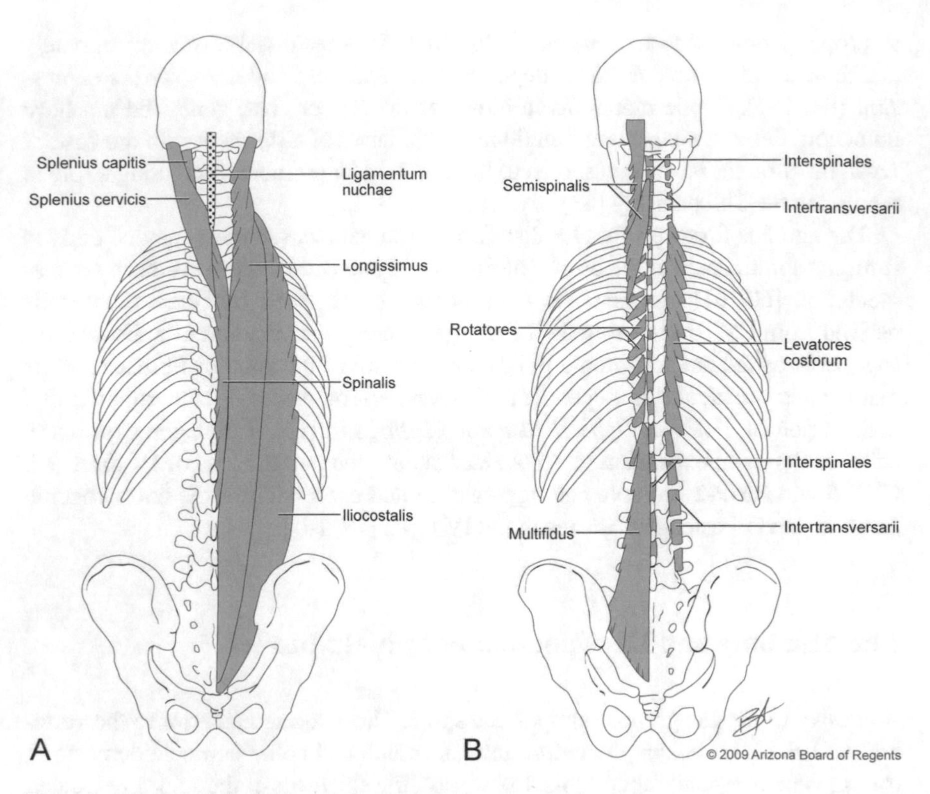

Fig. 1.8 Muscles of the back. (**A**) On the left, the superficial splenius muscles; on the right, the erector spinae muscles, including iliocostalis, longissimus, and spinalis. (**B**) On the left, the transversospinalis muscles, including semispinalis, multifidus, and rotatores; on the right, the levatores costarum, intertransversarii, and interspinales muscles (Drawing by Brent Adrian)

(Fig. 1.8A). Bilateral contraction of splenius capitis and cervicis extends the head and cervical spine while unilateral contraction laterally flexes and rotates the neck to the ipsilateral side.

Lying deep to the splenius layer, the erector spinae consist of three longitudinal columns of muscle (Fig. 1.8A). These muscles arise via a common tendon from the iliac crest, sacrum, and lumbar spinous processes. From lateral to medial, the columns include (1) iliocostalis, which attaches to the ribs and cervical transverse processes; (2) longissimus, which attaches to the ribs, thoracic and cervical transverse processes, and mastoid process; and (3) spinalis, which spans adjacent spinous processes and terminates on the occipital bone. Unilateral contraction of the erector spinae muscles laterally flexes and rotates the spine to the ipsilateral side while bilateral contraction extends the spine.

The transversospinalis muscles lie deep to the erector spinae. These muscles occupy the region between the transverse and spinous processes and include the semispinalis, multifidus, and rotatores muscles (Fig. 1.8B). The semispinalis muscles are located in the thoracic and cervical regions, while the rotatores are prominent in the thoracic region. In contrast, the multifidus extends along the length of the spine but is most developed in the lumbar region. Unilateral contraction of the transversospinalis muscles rotates the spine to the contralateral side, while bilateral contraction extends the spine. These muscles also stabilize adjacent vertebrae and may have a proprioceptive function [23, 100].

Deep to the erector spinae are the levatores costarum, intertransversarii, interspinales, and the muscles of the suboccipital triangle (Fig. 1.8B). The levatores costarum are located between the transverse processes and the ribs and act as accessory muscles of respiration. The intertransversarii and the interspinales span the transverse and spinous processes, respectively, and help stabilize the spine. Finally, among the muscles of the suboccipital triangle, the rectus capitis posterior major and minor and the superior oblique extend the atlanto-occipital joints, while the inferior oblique rotates the atlanto-axial joints.

The extensor muscles of the spine may contribute to either the initiation or the progression of scoliotic curves [30, 50, 85, 92, 157]. Asymmetry of the spinal extensors, especially multifidus, has been reported in individuals with idiopathic scoliosis, including different degrees of hypertrophy, atrophy, fiber type distribution, centralization of nuclei, electromyographic activity, and disruption of sarcotubular and myofibrillar elements [1, 7, 22, 24, 30, 32, 50, 52, 71, 85, 92, 118, 120, 125, 140, 165, 168, 169]. Whether these conditions are responsible for the development of idiopathic scoliosis, its progression, or both, is unclear.

Development of Spinal Muscles

The spinal muscles that function to stabilize and extend the vertebral column are derived from the dorsal half of the myotome, from the occipital, thoracic, lumbar, and sacral somites. The origins of spinal muscles lie within a highly mitogenic myogenic progenitor cell (MPC) population located in the dorsomedial margin of the dermomyotome. These cells migrate subjacently to a space between the dermomyotome and the sclerotome where they exit the cell cycle and differentiate into mononucleated myocytes (Fig. 1.6, [43, 105]). The myotome expands along both the medial/lateral and dorsal/ventral axes by successive waves of MPC migration from the dermomyotome [42, 43, 66, 106]. This is followed by fusion of the myocytes into the multinucleated myotubes and morphogenic remodeling into the pattern of the adult spinal muscles [158].

The genetic basis of skeletal muscle development has been an area of intense study. The myogenic bHLH transcription factor family, including MyoD (*Myod1*), myf-5 (*Myf5*), myogenin (*Myog*), and MRF4 (*Myf6*), has been shown to be essential to initiate and maintain the myogenic program in cells fated to the myogenic lineage. The phenotypes of individual and compound null mutants reveal that these factors can be split into a specification subclass (myf-5 and MyoD) and a differentiation subclass (myogenin and MRF4). Interaction between the myogenic bHLH factors and members of the myocyte enhancer factor-2 (MEF2) family of MADS-box transcription factors enhance muscle differentiation by increasing affinity of DNA binding and expanding the number of target genes that can be activated (reviewed in [4, 97]). The activity of Mef-2 and the myogenic factors are controlled in part by their association with chromatin remodeling proteins histone acetyltransferases (HATs) and histone deacetylases (HDAC) that promote and repress muscle-specific transcription, respectively. Calcium/calmodulin-dependent protein kinase (CaMK)-dependent phosphorylation of HDAC5 leads to its dissociation with MEF2 and transport out of the nucleus [89, 90]. Acetylation of MyoD and myf-5 through p300 or PCAF increases affinity of the transcription factors for its DNA target and promotes transcription of myogenin and MRF4 as well as induces cell cycle arrest [109, 115, 126].

Specification of MPCs within the somite fated to become the epaxial muscles is dependent on paracrine factors secreted by adjacent tissues. These signals direct the competence of the cells to initiate the myogenic program and promote the amplification of these committed progenitor cells in the dorsal/medial lip of the dermomyotome. Because of its role in specification, initiating *Myf5* transcription has been used as a readout of specification. A combination of sonic hedgehog (*Shh*) secreted from the notochord and Wnts from the dorsal neural tube and surface ectoderm are implicated in this process [14, 39, 117]. Based on explant experiments, *Wnt1* is able to induce the transcription of *Myf5* [145]. The activity is transduced by frizzled receptors 1 and 6 through the canonical β-catenin pathway [12]. The role of Shh in specification was first predicted by the absence of *Myf5* expression in the region of the epaxial myotome in *Shh* null embryos [13]. Further, mutations in Gli transcription factors, which transduce Shh signaling, also display a deficit in *Myf5* expression [87]. Consistent with these observations, the *Myf5* epaxial enhancer is dependent on a consensus binding sequence for Gli transcription factors and consensus binding sequence for Tcf/Lef, the β-catenin cofactor [12, 142, 153].

Though the cellular events associated with establishing the early muscle masses are now well described, as well as the genetic basis for muscle differentiation, less is known about subsequent events associated with establishing individual muscle groups from these masses. Embryonic muscles experience rapid growth, while the early muscle masses in the dorsal body wall, limb, hypoglossal chord, and head undergo several morphological processes (splitting, fusion, directional growth, and movement) in order to establish the appropriate shape, position, and fiber orientation of neonatal muscle. Further, they must coordinate with the growth and differentiation of tendons, ligaments, connective tissue, and skeletal elements to establish the appropriate origin and insertion sites on the bones. Patterning of muscle is dependent on innervation [164] and extrinsic signals from the surrounding tissue [61, 68]. This is mediated at least in part through mesodermal cells expressing *Tcf4* [68] and both intrinsic and extrinsic cues from members of the Hox gene family [2, 6]. In addition, the occurrence of defects in the multifidus muscles of mice with *Lfng* and *Dll3* mutations suggests a previously unappreciated role for Notch signaling in the patterning of the spinal muscles [51]. However, a clear understanding of the combination of local and global signals that directs individual and functional groups of muscles remains poorly understood.

Tendon Development

Tendons consist of fibroblast-like cells, called tenocytes, encased in a complex of collagen fibrils comprised of type I, III, IV, V, and VI collagen, Tenomodulin, and sulfated proteoglycans, including decorin, biglycan, fibromodulin, lumican, and aggrecan [133]. The embryonic formation of tendons occurs through the alignment of tenocytes along a linear track, followed by the deposition of the collagen fibrils. Tenocytes in mature tendon are thought to be in a non-proliferative quiescent state, with additional growth associated with an increase in collagen production [67]. Repair of tendons appears to be dependent on a localized stem cell population predicting an approach to injury repair similar to skeletal muscle [141].

The coordinated development of tendons along with muscle and skeletal elements is essential to the proper functioning of the musculoskeletal system [18]. However, the cellular origins of tendons and the regulator pathways that control their specification and differentiation are poorly understood. The identification of the bHLH transcription factor, scleraxis, as a tendon-specific marker accelerated research in this area [155]. Consistent with its intimate relationship to the epaxial muscles and vertebrae, the axial tendon is derived from a subdomain of the somite referred to as the syndetome, which is located between the myotome and sclerotome (Fig. 1.7) [17, 132]. The syndetomal cells are derived from an interaction between the sclerotome and myotome. Expression of *Fgf4* and *Fgf8* in the myotome is both necessary and sufficient for scleraxis expression in sclerotomal cells in the future syndetome region [17, 18]. Within the sclerotomal cells, the FGF induces an ERK MAP kinase-mediated cascade that requires activation of the ETS transcription factor, *Etv4/Pea3* [16, 136]. It appears that there are also inhibitory signals generated from the sclerotome that limit the size of the syndetome. Overexpression of *Pax1* reduces the scleraxis expression domain in the sclerotome, a compound mouse mutation in *Sox5/Sox6* lead to an expansion of the scleraxis-expressing domain [18].

Several regulators have been identified that are essential for tenocyte differentiation as well as tendon maturation and maintenance leading to a simple model for tendogenesis. TGFβ and FGF signaling specify tenocytes from mesenchymal progenitors in part by the induction of the bHLH transcription factor, Scleraxis (*Scx*) [48, 114]. This is followed by the expression of Mohawk (*Mkx*) and early growth response factors 1 and 2 (*Egr1* and *Egr2*) in tenocytes. These genes are maintained in the tendons after birth, while *Scx* transcription levels diminish [133]. This predicts distinct functions for the three transcription factors that ultimately promote the expression of elements of the collagen fibrils associated with tendon development and adult maintenance and repair [5].

Targeted null mutations in mice have been leveraged extensively to determine the function of these genes in tendogenesis. Inactivation of Scx (*Scx−/−*) resulted in a significant loss of tendons in the limbs, trunk, and tail [102]. However, this did not eliminate all tendons, suggesting the presence of additional factors that are differentially used in tenocyte differentiation. In contrast, *Mkx−/−* mice displayed a reduction in tendon mass through hypoplastic tendon fibers but no reduction in tenocyte numbers [102]. This was recapitulated in rats, where *Mkx* was inactivated using the CRISPR-Cas9 system, suggesting an essential role for the gene in tendon maturation [143]. *Egr1* and *Egr2* mutations lead to a reduction of collagen fibrils and the expression of *Scx* and *Mkx* in embryonic tendons consistent with providing positive feedback in the tendon signaling cassette. It is important to note that none of these mutations lead to a complete ablation of tendon development. This predicts functional redundancy or the existence of additional regulators that have not yet been identified.

In adults, tendons participate in homeostatic sensing that matches force-transmission capacity to mechanical load through a mechanical sensing system. This leads to the differentiation of mesenchymal stem cells associated with the tendon to tenocytes [69]. This appears to recapitulate the embryonic signaling pathway, as it requires *Scx*, *Mkx*, and *Egr1* [54, 69, 93]. The general transcription factor II-I repeat domain-containing protein 1 (*Gtf2ird1*) has been found to be important in mechanical sensing. In response to stretching, *Gtf2ird1* translocates to the nucleus from the cytoplasm where it induces *Mkx* expression [69]. Interestingly, extreme stretching leads to tendon damage and a reduction of tendon-specific gene expression and an increase in osteogenic and chondrogenic gene markers [93]. This can be recapitulated under conditions of mild stress, including the expression of *Sox6*, *Sox9*, and aggrecan, by inactivation of *Mkx* [143]. This predicts that *Mkx* plays the dual role of promoting tendon differentiation and preventing chondrogenesis.

Summary

The vertebral column, spinal musculature, and associated tendons arise from paraxial mesoderm which undergoes segmentation in a rostral to caudal direction to form pairs of somites. Each somite differentiates into four cell lineage-specific compartments that contribute to the spine, including the sclerotome (vertebrae and ribs), myotome (skeletal muscle), dermomyotome (dermis and skeletal muscle progenitor cells), and syndetome (tendons). The timing of somite formation and the determination of the site of boundary formation are established by the interactions between the Notch, Wnt, and FGF signaling pathways. In this chapter, we focused on three essential aspects of somite formation and patterning, including the establishment of the intersomitic boundary, somite epithelialization, and rostral/caudal polarity, and the subsequent development of the vertebrae, IV discs, and associated spinal muscles and tendons.

The vertebrae are derived from the sclerotome compartment of the somites. The transition from sclerotome to vertebrae can be divided into distinct processes for the ventral structures (vertebral body and intervertebral discs) and dorsal neural arch structures (pedicles, laminae, spinous and transverse processes) based on both cell origin and genetic regulation. An additional layer of regulation, via members of the Hox transcription factor family, is required to confer the distinctive regional characteristics of the cervical, thoracic, lumbar, sacral, and caudal vertebrae. The IV discs, comprised of a nucleus pulposus, annulus fibrosus, and cartilage end plates, were originally thought to be derived solely from the sclerotome, but the nucleus pulposus has been shown to be derived from the notochord. As a result, we must now invoke a more complicated model for the development of the IV discs that requires the coordination of multiple independent signaling pathways such as the *Shh* and TGF-beta superfamily.

The spinal muscles that function to stabilize and extend the vertebral column are derived from a highly mitogenic myogenic progenitor cell population located in the dorsomedial margin of the dermomyotome. These cells migrate to a space between the dermomyotome and the sclerotome where they exit the cell cycle and differentiate into mononucleated myocytes, forming the myotome. Though the cellular events associated with establishing the early muscle masses are now well described, as well as the genetic basis for muscle differentiation, less is known about the events leading to the development of individual and functional groups of spinal muscles. The tendons associated with these muscles are derived from a subdomain of the somite referred to as the syndetome, which is located between the myotome and sclerotome. The cellular origins of tendons and the regulator pathways that control their specification and differentiation are also poorly understood, although recent work in this area has identified several regulators, such as TGFβ and FGF signaling, as essential for tenocyte differentiation as well as tendon maturation and maintenance.

Acknowledgments We would like to thank Brent Adrian for preparing Figs. 1.1, 1.3, 1.4, 1.5, and 1.8.

References

1. Alexander MA, Season EH. Idiopathic scoliosis: an electromyographic study. Arch Phys Med Rehabil. 1978;59:314–5.
2. Alvares LE, Schubert FR, Thorpe C, Mootoosamy RC, Cheng L, Parkyn G, et al. Intrinsic, Hox-dependent cues determine the fate of skeletal muscle precursors. Dev Cell. 2003;5:379–90.
3. Aoyama H, Asamoto K. The developmental fate of the rostral/caudal half of a somite for vertebra and rib formation: experimental confirmation of the resegmentation theory using chick-quail chimeras. Mech Dev. 2000;99:71–82.
4. Arnold HH, Braun T. Genetics of muscle determination and development. Curr Top Dev Biol. 2000;48:129–64.
5. Asahara H, Inui M, Lotz MK. Tendon and ligaments: connecting development biology to musculoskeletal disease pathogenesis. J Bone Miner Res. 2017;32:1773–82.
6. Ashby P, Chinnah T, Zakany J, Duboule D, Tickle C. Muscle and tendon pattern is altered independently of skeletal pattern in HoxD mutant limbs. J Anat. 2002;201:422.
7. Avikainen VJ, Rezasoltani A, Kauhanen HA. Asymmetry of paraspinal EMG-time characteristics in idiopathic scoliosis. J Spinal Disord. 1999;12:61–7.
8. Baffi MO, Moran MA, Serra R. Tgfbr2 regulates the maintenance of boundaries in the axial skeleton. Dev Biol. 2006;296:363–74.

9. Barrallo-Gimeno A, Nieto MA. The snail genes as inducers of cell movement and survival: implications in development and cancer. Development. 2005;132:3151–61.
10. Batlle E, Sancho E, Franci C, Dominguez D, Monfar M, Baulida J, et al. The transcription factor snail is a repressor of E-cadherin gene expression in epithelial tumour cells. Nat Cell Biol. 2000;2:84–9.
11. Berkes CA, Bergstrom DA, Penn BH, Seaver KJ, Knoepfler PS, Tapscott SJ. Pbx marks genes for activation by MyoD indicating a role for a homeodomain protein in establishing myogenic potential. Mol Cell. 2004;14:465–77.
12. Borello U, Berarducci B, Murphy P, Bajard L, Buffa V, Piccolo S, et al. The Wnt/beta-catenin pathway regulates Gli-mediated Myf5 expression during somitogenesis. Development. 2006;133:3723–32.
13. Borycki AG, Brunk B, Tajbakhsh S, Buckingham M, Chiang C, Emerson CP Jr. Sonic hedgehog controls epaxial muscle determination through Myf5 activation. Development. 1999;126:4053–63.
14. Borycki A, Brown AM, Emerson CP Jr. Shh and Wnt signaling pathways converge to control Gli gene activation in avian somites. Development. 2000;127:2075–87.
15. Brand-Saberi B, Christ B. Evolution and development of distinct cell lineages derived from somites. Curr Top Dev Biol. 2000;48:1–42.
16. Brent AE, Tabin CJ. FGF acts directly on the somitic tendon progenitors through the Ets transcription factors Pea3 and Erm to regulate scleraxis expression. Development. 2004;131:3885–96.
17. Brent AE, Schweitzer R, Tabin CJ. A somitic compartment of tendon progenitors. Cell. 2003;113:235–48.
18. Brent AE, Braun T, Tabin CJ. Genetic analysis of interactions between the somitic muscle, cartilage and tendon cell lineages during mouse development. Development. 2005;132:515–28.
19. Buchberger A, Seidl K, Klein C, Eberhardt H, Arnold HH. cMeso-1, a novel bHLH transcription factor, is involved in somite formation in chicken embryos. Dev Biol. 1998;199:201–15.
20. Burgess R, Cserjesi P, Ligon KL, Olson EN. Paraxis: a basic helix-loop-helix protein expressed in paraxial mesoderm and developing somites. Dev Biol. 1995;168:296–306.
21. Burgess R, Rawls A, Brown D, Bradley A, Olson EN. Requirement of the paraxis gene for somite formation and musculoskeletal patterning. Nature. 1996;384:570–3.
22. Butterworth TR, James C. Electromyographic studies in idiopathic scoliosis. South Med J. 1969;62:1008–10.
23. Buxton DF, Peck D. Neuromuscular spindles relative to joint movement complexities. Clin Anat. 1989;2:211–24.
24. Bylund P, Jansson E, Dahlberg E, Eriksson E. Muscle fiber types in thoracic erector spinae muscles. Clin Orthop. 1987;214:222–8.
25. Cailliet R. Low back pain syndrome. 4th ed. Philadelphia: FA Davis Company; 1988.
26. Cano A, Perez-Moreno MA, Rodrigo I, Locascio A, Blanco MJ, del Barrio MG, et al. The transcription factor snail controls epithelial-mesenchymal transitions by repressing E-cadherin expression. Nat Cell Biol. 2000;2:76–83.
27. Capellini TD, Di Giacomo G, Salsi V, Brendolan A, Ferretti E, Srivastava D, et al. Pbx1/Pbx2 requirement for distal limb patterning is mediated by the hierarchical control of Hox gene spatial distribution and Shh expression. Development. 2006;133:2263–73.
28. Capellini TD, Zewdu R, Di Giacomo G, Asciutti S, Kugler JE, Di Gregorio A, et al. Pbx1/Pbx2 govern axial skeletal development by controlling Polycomb and Hox in mesoderm and Pax1/Pax9 in sclerotome. Dev Biol. 2008;321:500–14.
29. Chal J, Guillot C, Pourquié O. PAPC couples the segmentation clock to somite morphogenesis by regulating N-cadherin-dependent adhesion. Development. 2017;144:664–76.
30. Chan YL, Cheng JCY, Guo X, King AD, Griffith JF, Metreweli C. MRI evaluation of multifidus muscles in adolescent idiopathic scoliosis. Pediatr Radiol. 1999;29:360–3.
31. Chen F, Greer J, Capecchi MR. Analysis of Hoxa7/Hoxb7 mutants suggests periodicity in the generation of different sets of vertebrae. Mech Dev. 1998;77:49–57.

32. Cheung J, Halbertsma JPK, Veldhuizen AG, Sluiter WJ, Maurits NM, Cool JC, et al. A preliminary study on electromyographic analysis of the paraspinal musculature in idiopathic scoliosis. Eur Spine J. 2005;14:130–7.
33. Chiang C, Litingtung Y, Lee E, Young KE, Corden JL, Westphal H, et al. Cyclopia and defective axial patterning in mice lacking Sonic hedgehog gene function. Nature. 1996;383:407–13.
34. Choi KS, Harfe BD. Hedgehog signaling is required for formation of the notochord sheath and patterning of nuclei pulposi within the intervertebral discs. Proc Natl Acad Sci U S A. 2011;108:9484–9.
35. Choi KS, Lee C, Harfe BD. Sonic hedgehog in the notochord is sufficient for patterning of the intervertebral discs. Mech Dev. 2012;129:255–62.
36. Christ B, Wilting J. From somites to vertebral column. Ann Anat. 1992;174:23–32.
37. Condie BG, Capecchi MR. Mice with targeted disruptions in the paralogous genes hoxa-3 and hoxd-3 reveal synergistic interactions. Science. 1994;370:304–7.
38. Correia KM, Conlon RA. Surface ectoderm is necessary for the morphogenesis of somites. Mech Dev. 2000;91:19–30.
39. Cossu G, Borello U. Wnt signaling and the activation of myogenesis in mammals. EMBO J. 1999;18:6867–72.
40. Dale JK, Malapert P, Chal J, Vilhais-Neto G, Maroto M, Johnson T, Jayasinghe S, Trainor P, Herrmann B, Pourquié O. Oscillations of the snail genes in the presomitic mesoderm coordinate segmental patterning and morphogenesis in vertebrate somitogenesis. Dev Cell. 2006;10:355–66.
41. de la Pompa JL, Wakeham A, Correia KM, Samper E, Brown S, Aguilera RJ, et al. Conservation of the Notch signaling pathway in mammalian neurogenesis. Development. 1997;124:1139–48.
42. Denetclaw WF Jr, Ordahl CP. The growth of the dermomyotome and formation of early myotome lineages in thoracolumbar somites of chicken embryos. Development. 2000;127:893–905.
43. Denetclaw WF Jr, Christ B, Ordahl CP. Location and growth of epaxial myotome precursor cells. Development. 1997;124:1601–10.
44. Dockter JL. Sclerotome induction and differentiation. Curr Top Dev Biol. 2000;48:77–127.
45. Duband JL, Dufour S, Hatta K, Takeichi M, Edelman GM, Thiery JP. Adhesion molecules during somitogenesis in the avian embryo. J Cell Biol. 1987;104:1361–74.
46. Dubrulle J, Pourquié O. Coupling segmentation to axis formation. Development. 2004;131:5783–93.
47. Dunwoodie SL, Henrique D, Harrison SM, Beddington RSP. Mouse *Dll3*: a novel divergent *Delta* gene which may complement the function of other Delta homologues during early pattern formation in the mouse embryo. Development. 1997;124:3065–76.
48. Eloy-Trinquet S, Wang H, Edom-Vovar F, Duprez D. Fgf signaling components are associated with muscles and tendons during limb development. Dev Dyn. 2009;238:1195–206.
49. Fan CM, Tessier-Lavigne M. Patterning of mammalian somites by surface ectoderm and notochord: evidence for sclerotome induction by a hedgehog homolog. Cell. 1994;79:1175–86.
50. Fidler MW, Jowett RL. Muscle imbalance in the aetiology of scoliosis. J Bone Joint Surg. 1976;58-B:200–1.
51. Fisher RE, Smith HF, Kusumi K, Tassone EE, Rawls A, Wilson-Rawls J. Mutations in the Notch pathway alter the patterning of multifidus. Anat Rec. 2012;295:32–9.
52. Ford DM, Bagnall KM, McFadden KD, Greenhill BJ, Raso VJ. Paraspinal muscle imbalance in adolescent idiopathic scoliosis. Spine. 1984;9:373–6.
53. Furumoto TA, Miura N, Akasaka T, Mizutanikoseki Y, Sudo H, Fukuda K, et al. Notochord-dependent expression of MFH1 and PAX1 cooperates maintain the proliferation of sclerotome cells during the vertebral column development. Dev Biol. 1999;210:15–29.
54. Gaut L, Robert N, Delalande A, Bonnin MA, Pichon C, Duprez D. EGR1 regulates transcription downstream of mechanical signals during tendon formation and healing. PLoS One. 2016;11:e0166237.

55. Geetha-Loganathan P, Nimmagadda S, Huang R, Christ B, Scaal M. Regulation of ectodermal Wnt6 expression by the neural tube is transduced by dermomyotomal Wnt11: a mechanism of dermomyotomal lip sustainment. Development. 2006;133:2897–904.
56. Goldstein RS, Kalcheim C. Determination of epithelial half-somites in skeletal morphogenesis. Development. 1992;116:441–5.
57. Henry CA, Hall LA, Burr Hille M, Solnica-Krezel L, Cooper MS. Somites in zebrafish doubly mutant for knypek and trilobite form without internal mesenchymal cells or compaction. Curr Biol. 2000;10:1063–6.
58. Horikawa K, Radice G, Takeichi M, Chisaka O. Adhesive subdivisions intrinsic to the epithelial somites. Dev Biol. 1999;215:182–9.
59. Hrabě de Angelis M, McIntyre J 2nd, Gossler A. Maintenance of somite borders in mice requires the Delta homologue DII1. Nature. 1997;386:717–21.
60. Huang R, Zhi Q, Neubuser A, Muller TS, Brand-Saberi B, Christ B, et al. Function of somite and somitocoele cells in the formation of the vertebral motion segment in avian embryos. Acta Anat. 1996;155:231–41.
61. Jacob HJ, Christ B. On the formation of muscular pattern in the chick limb. In: Teratology of the limbs. Berlin: Walter de Gruyter and Co.; 1988. p. 89–97.
62. Jacob M, Jacob JH, Christ B. The early differentiation of the perinotochordal connective tissue. A scanning and transmission electron microscopic study on chick embryos. Experientia. 1975;31:1083–6.
63. Jeong Y, Epstein DJ. Distinct regulators of Shh transcription in the floor plate and notochord indicate separate origins for these tissues in the mouse node. Development. 2003;130:3891–902.
64. Jiang YJ, Aerne BL, Smithers L, Haddon C, Ish-Horowicz D, Lewis J. Notch signaling and the synchronization of the somite segmentation clock. Nature. 2000;408:475–9.
65. Johnson J, Rhee J, Parsons SM, Brown D, Olson EN, Rawls A. The anterior/posterior polarity of somites is disrupted in paraxis-deficient mice. Dev Biol. 2001;229:176–87.
66. Kahane N, Cinnamon Y, Kalcheim C. The cellular mechanism by which the dermomyotome contributes to the second wave of myotome development. Development. 1998;125:4259–71.
67. Kalson NS, Lu Y, Taylor SH, Starborg T, Homes DF, Kadler KE. A structure-based extracellular matrix expansion mechanism of fibrous tissue growth. elife. 2015;4:e05958.
68. Kardon G, Harfe BD, Tabin CT. A Tcf4-positive mesodermal population provides a prepattern for vertebrate limb muscle patterning. Dev Cell. 2015;5:937–44.
69. Kayama T, Mori M, Ito Y, Matsushima T, Nakamichi R, Suzuki H, et al. Gtf2ird1-dependent Mohawk expression regulates Mechanosensing properties of the tendon. Mol Cell Biol. 2016;36:1297–309.
70. Keynes RJ, Stern CD. Mechanisms of vertebrate segmentation. Development. 1988;103:413–29.
71. Khosla S, Tredwell SJ, Day B, Shinn SL, Ovalle WK. An ultrastructural study of multifidus muscle in progressive idiopathic scoliosis-changes resulting from a sarcolemmal defect of the myotendinous junction. J Neurol Sci. 1980;46:13–31.
72. Kim SY, Paylor SW, Magnuson T, Schumacher A. Juxtaposed Polycomb complexes co-regulate vertebral identity. Development. 2006;133:4957–68.
73. Koizumi K, Nakajima M, Yuasa S, Saga Y, Sakai T, Kuriyama T, et al. The role of presenilin 1 during somite segmentation. Development. 2001;128:1391–402.
74. Kulesa PM, Fraser SE. Cell dynamics during somite boundary formation revealed by time-lapse analysis. Science. 2002;298:991–5.
75. Kulesa PM, Schnell S, Rudloff S, Baker RE, Maini PK. From segment to somite: segmentation epithelialization analyzed within quantitative frameworks. Dev Dyn. 2007;236:1392–402.
76. Kusumi K, Sun ES, Kerrebrock AW, Bronson RT, Chi DC, Bulotsky MS, et al. The mouse pudgy mutation disrupts Delta homologue Dll3 and initiation of early somite boundaries. Nat Genet. 1998;19:274–8.
77. Lewis EB. A gene complex controlling segmentation in Drosophila. Nature. 1978;276:565–70.

78. Li J, Yoon ST, Hutton WC. Effect of bone morphogenetic protein-2 (BMP-2) on matrix production, other BMPs, and BMP receptors in rat intervertebral disc cells. J Spinal Disord Tech. 2004;17:423–8.
79. Linask KK, Ludwig C, Han MD, Liu X, Radice GL, Knudsen KA. N-cadherin/catenin-mediated morphoregulation of somite formation. Dev Biol. 1998;202:85–102.
80. Linker C, Lesbros C, Gros J, Burrus LW, Rawls A, Marcelle C. Beta-catenin-dependent Wnt signalling controls the epithelial organisation of somites through the activation of paraxis. Development. 2005;132:3895–905.
81. Mackie EJ, Ahmed YA, Tatarczuch L, Chen KS, Mirams M. Endochondral ossification: how cartilage is converted into bone in the developing skeleton. Int J Biochem Cell Biol. 2008;40:46–62.
82. Maconochie MK, Nonchev S, Studer M, Chan SK, Popperl H, Sham MH, et al. Cross-regulation in the mouse HoxB complex: the expression of Hoxb2 in rhombomere 4 is regulated by Hoxb1. Genes Dev. 1997;11:1885–95.
83. Maier JA, Lo Y, Harfe BD. Foxa1 and Foxa2 are required for formation of the intervertebral discs. PLoS One. 2013;8:e55528.
84. Mankoo BS, Skuntz S, Harrigan I, Grigorieva E, Candia A, Wright CV, et al. The concerted action of Meox homeobox genes is required upstream of genetic pathways essential for the formation, patterning and differentiation of somites. Development. 2003;130:4655–64.
85. Mannion AF, Meier M, Grob D, Müntener M. Paraspinal muscle fibre type alterations associated with scoliosis: an old problem revisited with new evidence. Eur Spine J. 1998;7:289–93.
86. Mansouri A, Pla P, Larue L, Gruss P. Pax3 acts cell autonomously in the neural tube and somites by controlling cell surface properties. Development. 2001;128:1995–2005.
87. McDermott A, Gustafsson M, Elsam T, Hui CC, Emerson CP Jr, Borycki AG. Gli2 and Gli3 have redundant and context-dependent function in skeletal muscle formation. Development. 2005;132:345–57.
88. McIntyre DM, Rakshit S, Yallowitz AR, Loken L, Jeannotte L, Capecchi MR, et al. *Hox* patterning of the vertebrate rib cage. Development. 2007;134:2981–9.
89. McKinsey TA, Zhang CL, Lu J, Olson EN. Signal-dependent nuclear export of a histone deacetylase regulates muscle differentiation. Nature. 2000;408:106–11.
90. McKinsey TA, Zhang CL, Olson EN. Control of muscle development by dueling HATs and HDACs. Curr Opin Genet Dev. 2001;11:497–504.
91. McMahon JA, Takada S, Zimmerman LB, McMahon AP. Noggin-mediated antagonism of BMP signaling is required for growth and patterning of the neural tube and somite. Genes Dev. 1998;12:1438–52.
92. Meier MP, Klein MP, Krebs D, Grob D, Müntener M. Fiber transformations in multifidus muscle of young patients with idiopathic scoliosis. Spine. 1997;22:2357–64.
93. Mendias CL, Gumucio JP, Bakhurin KI, Lynch EB, Brooks SV. Physiological loading of tendons induces scleraxis expression in epitenon fibroblasts. J Orthop Res. 2012;30:606–12.
94. Mercer SR, Bogduk N. Clinical anatomy of ligamentum nuchae. Clin Anat. 2003;16:484–93.
95. Mittapalli VR, Huang R, Patel K, Christ B, Scaal M. Arthrotome: a specific joint forming compartment in the avian somite. Dev Dyn. 2005;234:48–53.
96. Moens CB, Selleri L. Hox cofactors in vertebrate development. Dev Biol. 2006;291:193–206.
97. Molkentin JD, Olson EN. Defining the regulatory networks for muscle development. Curr Opin Genet Dev. 1996;6:445–53.
98. Monsoro-Burq AH, Bontoux M, Teillet MA, Le Douarin NM. Heterogeneity in the development of the vertebra. Proc Natl Acad Sci U S A. 1994;91:10435–9.
99. Monsoro-Burq AH, Duprez D, Watanabe Y, Bontoux M, Vincent C, Brickell P, et al. The role of bone morphogenetic proteins in vertebral development. Development. 1996;122:3607–16.
100. Moore KL, Dalley AF. Clinically oriented anatomy. Baltimore: Lippincott Williams and Wilkins; 2006.
101. Morimoto M, Sasaki N, Oginuma M, Kiso M, Igarashi K, Aizaki K, et al. The negative regulation of Mesp2 by mouse Ripply2 is required to establish the rostro-caudal patterning within a somite. Development. 2007;134:1561–9.

102. Murchison ND, Price BA, Conner DA, Keene DR, Olson EN, Tabin CJ, et al. Regulation of tendon differentiation by scleraxis distinguishes force-transmitting tendons from muscle-anchoring tendons. Development. 2007;134:2697–708.
103. Nakaya Y, Kuroda S, Katagiri YT, Kaibuchi K, Takahashi Y. Mesenchymal-epithelial transition during somitic segmentation is regulated by differential roles of Cdc42 and Rac1. Dev Cell. 2004;7:425–38.
104. Oka C, Nakano T, Wakeham A, de la Pompa JL, Mori C, Sakai T, et al. Disruption of the mouse *RBP-J kappa* gene results in early embryonic death. Development. 1995;121:3291–301.
105. Ordahl CP, Le Douarin NM. Two myogenic lineages within the developing somite. Development. 1992;114:339–53.
106. Ordahl CP, Berdougo E, Venters SJ, Denetclaw WF Jr. The dermomyotome dorsomedial lip drives growth and morphogenesis of both the primary myotome and dermomyotome epithelium. Development. 2001;128:1731–44.
107. Paavola LG, Wilson DB, Center EM. Histochemistry of the developing notochord, perichordal sheath and vertebrae in Danforth's short-tail (sd) and normal C57BL/6 mice. J Embryol Exp Morphol. 1980;55:227–45.
108. Palmeirim I, Dubrulle J, Henrique D, Ish-Horowicz D, Pourquié O. Uncoupling segmentation and somitogenesis in the chick presomitic mesoderm. Dev Genet. 1998;23:77–85.
109. Peschiaroli A, Figliola R, Coltella L, Strom A, Valentini A, D'Agnano I, et al. MyoD induces apoptosis in the absence of RB function through a p21(WAF1)-dependent re-localization of cyclin/cdk complexes to the nucleus. Oncogene. 2002;21:8114–27.
110. Peters H, Doll U, Niessing J. Differential expression of the chicken Pax-1 and Pax-9 gene: in situ hybridization and immunohistochemical analysis. Dev Dyn. 1995;203:1–16.
111. Peters H, Wilm B, Sakai N, Imai K, Maas R, Balling R. Pax1 and Pax9 synergistically regulate vertebral column development. Development. 1999;126:5399–408.
112. Popperl H, Bienz M, Studer M, Chan SK, Aparicio S, Brenner S, et al. Segmental expression of Hoxb-1 is controlled by a highly conserved autoregulatory loop dependent upon exd/pbx. Cell. 1995;81:1031–42.
113. Pourquie O, Coltey M, Teillet MA, Ordahl C, Le Douarin M. Control of dorsoventral patterning of somitic derivatives by notochord and floor plate. Proc Natl Acad Sci U S A. 1993;90:5242–6.
114. Pryce B, Watson SS, Murchison ND, Staverosky JA, Dunker N, Schweitzer R. Recruitment and maintenance of tendon progenitors by TGFbeta signaling are essential for tendon formation. Development. 2009;136:1351–61.
115. Puri PL, Sartorelli V, Yang XJ, Hamamori Y, Ogryzko VV, Howard BH, et al. Differential roles of p300 and PCAF acetyltransferases in muscle differentiation. Mol Cell. 1997;1:35–45.
116. Radice GL, Rayburn H, Matsunami H, Knudsen KA, Takeichi M, Hynes RO. Developmental defects in mouse embryos lacking N-cadherin. Dev Biol. 1997;181:64–78.
117. Reshef R, Maroto M, Lassar AB. Regulation of dorsal somitic cell fates: BMPs and Noggin control the timing and pattern of myogenic regulator expression. Genes Dev. 1998;12:290–303.
118. Reuber M, Schultz A, McNeill T, Spencer D. Trunk muscle myoelectric activities in idiopathic scoliosis. Spine. 1983;8:447–56.
119. Rhee J, Takahashi Y, Saga Y, Wilson-Rawls J, Rawls A. The protocadherin papc is involved in the organization of the epithelium along the segmental border during mouse somitogenesis. Dev Biol. 2003;254:248–61.
120. Riddle HF, Roaf R. Muscle imbalance in the causation of scoliosis. Lancet. 1955;268:1245–7.
121. Rodrigo I, Hill RE, Balling R, Münsterberg A, Imai K. Pax1 and Pax9 activate Bapx1 to induce chondrogenic differentiation in the sclerotome. Development. 2003;130:473–82.
122. Rodrigo I, Bovolenta P, Mankoo BS, Imai K. Meox homeodomain proteins are required for Bapx1 expression in the sclerotome and activate its transcription by direct binding to its promoter. Mol Cell Biol. 2004;24:2757–566.

123. Rowton M, Ramos P, Anderson DM, Rhee JM, Cunliffe HE, Rawls A. Regulation of mesenchymal-to-epithelial transition by *Paraxis* during somitogenesis. Dev Dyn. 2013;242:1332–44.
124. Saga Y, Hata N, Koseki H, Taketo MM. Mesp2: a novel mouse gene expressed in the presegmented mesoderm and essential for segmentation initiation. Genes Dev. 1997;11:1827–39.
125. Sahgal V, Shah A, Flanagan N, Schaffer M, Kane W, Subramani V, et al. Morphologic and morphometric studies of muscle in idiopathic scoliosis. Acta Orthop. 1983;54:242–51.
126. Sartorelli V, Puri PL, Hamamori Y, Ogryzko V, Chung G, Nakatani Y, et al. Acetylation of MyoD directed by PCAF is necessary for the execution of the muscle program. Mol Cell. 1999;4:725–34.
127. Sato Y, Takahashi Y. A novel signal induces a segmentation fissure by acting in a ventral-to-dorsal direction in the presomitic mesoderm. Dev Biol. 2005;282:183–91.
128. Sato Y, Yasuda K, Takahashi Y. Morphological boundary forms by a novel inductive event mediated by Lunatic fringe and Notch during somitic segmentation. Development. 2002;129:3633–44.
129. Schmidt C, Stoeckelhuber M, McKinnell I, Putz R, Christ B, Patel K. Wnt 6 regulates the epithelialisation process of the segmental plate mesoderm leading to somite formation. Dev Biol. 2004;271:198–209.
130. Schubert FR, Tremblay P, Mansouri A, Faisst AM, Kammandel B, Lumsden A, et al. Early mesodermal phenotypes in splotch suggest a role for Pax3 in the formation of epithelial somites. Dev Dyn. 2001;222:506–21.
131. Schuster-Gossler K, Harris B, Johnson R, Serth J, Gossler A. Notch signalling in the paraxial mesoderm is most sensitive to reduced Pofut1 levels during early mouse development. BMC Dev Biol. 2009;9:6.
132. Schweitzer R, Chyung JH, Murtaugh LC, Brent AE, Rosen V, Olson EN, et al. Analysis of the tendon cell fate using Scleraxis, a specific marker for tendons and ligaments. Development. 2001;128:3855–66.
133. Schweitzer R, Zelzer E, Volk T. Connecting muscles to tendons: tendons and musculoskeletal development in flies and vertebrates. Development. 2010;137(17):2807.
134. Sivan SS, Hayes AJ, Wachtel E, Caterson B, Merkher Y, Maroudas A, et al. Biochemical composition and turnover of the extracellular matrix of the normal and degenerate intervertebral disc. Eur Spine J. 2014;23(Suppl 3):S344–53.
135. Skuntz S, Mankoo B, Nguyen MT, Hustert E, Nakayama A, Tournier-Lasserve E, et al. Lack of the mesodermal homeodomain protein MEOX1 disrupts sclerotome polarity and leads to a remodeling of the cranio-cervical joints of the axial skeleton. Dev Biol. 2009;332:383–95.
136. Smith TG, Sweetman D, Patterson M, Keyse SM, Münsterberg A. Feedback interactions between MKP3 and ERK MAP kinase control scleraxis expression and the specification of rib progenitors in the developing chick somite. Development. 2005;132:1305–14.
137. Smits P, Lefebvre V. Sox5 and Sox6 are required for notochord extracellular matrix sheath formation, notochord cell survival and development of the nucleus pulposus of intervertebral discs. Development. 2003;130:1135–48.
138. Sosić D, Brand-Saberi B, Schmidt C, Christ B, Olson EN. Regulation of paraxis expression and somite formation by ectoderm- and neural tube-derived signals. Dev Biol. 1997;185:229–43.
139. Sparrow DB, Chapman G, Turnpenny PD. Dunwoodie SL disruption of the somitic molecular clock causes abnormal vertebral segmentation. Birth Defects Res C Embryo Today. 2007;81:93–110.
140. Spencer GS, Zorab PA. Spinal muscle in scoliosis. Part 1: histology and histochemistry. J Neurol Sci. 1976;30:127–42.
141. Steinert AF, Kunz M, Prager P, Barthel T, Jakob F, Nöth U, et al. Mesenchymal stem cell characteristics of human anterior cruciate ligament outgrowth cells. Tissue Eng Part A. 2011;17:1375–88.

142. Summerbell D, Ashby PR, Coutelle O, Cox D, Yee S, Rigby PW. The expression of Myf5 in the developing mouse embryo is controlled by discrete and dispersed enhancers specific for particular populations of skeletal muscle precursors. Development. 2000;127:3745–57.
143. Suzuki H, Ito Y, Shinohara M, Yamashita S, Ichinose S, Kishida A, et al. Gene targeting of the transcription factor Mohawk in rats causes heterotopic ossification of Achilles tendon via failed tenogenesis. Proc Natl Acad Sci U S A. 2016;113:7840–5.
144. Swiatek PJ, Lindsell CE, del Amo FF, Weinmaster G, Gridley T. Notch1 is essential for post-implantation development in mice. Genes Dev. 1994;8:707–19.
145. Tajbakhsh S, Borello U, Vivarelli E, Kelly R, Papkoff J, Duprez D, et al. Differential activation of Myf5 and MyoD by different Wnts in explants of mouse paraxial mesoderm and the later activation of myogenesis in the absence of Myf5. Development. 1998;125:4155–62.
146. Takahashi Y, Sato Y. Somitogenesis as a model to study the formation of morphological boundaries and cell epithelialization. Develop Growth Differ. 2008;50:S149–55.
147. Takahashi Y, Koizumi K, Takagi A, Kitajima S, Inoue T, Koseki H, et al. Mesp2 initiates somite segmentation through the Notch signalling pathway. Nat Genet. 2000;25:390–6.
148. Takahashi Y, Inoue T, Gossler A, Saga Y. Feedback loops comprising Dll1, Dll3 and Mesp2, and differential involvement of Psen1 are essential for rostrocaudal patterning of somites. Development. 2003;130:4259–68.
149. Takahashi Y, Takagi A, Hiraoka S, Koseki H, Kanno J, Rawls A, et al. Transcription factors Mesp2 and Paraxis have critical roles in axial musculoskeletal formation. Dev Dyn. 2007;236:1484–94.
150. Takimoto A, Mohri H, Kokubu C, Hiraki Y, Shukunami C. Pax1 acts as a negative regulator of chondrocyte maturation. Exp Cell Res. 2013;319:3128–39.
151. Tam PP, Trainor PA. Specification and segmentation of the paraxial mesoderm. Anat Embryol. 1994;189:275–305.
152. Tanaka M, Tickle C. Tbx18 and boundary formation in chick somite and wing development. Dev Biol. 2004;268:470–80.
153. Teboul L, Summerbell D, Rigby PW. The initial somitic phase of Myf5 expression requires neither Shh signaling nor Gli regulation. Genes Dev. 2003;17:2870–4.
154. Tonegawa A, Funayama N, Ueno N, Takahashi Y. Mesodermal subdivision along the mediolateral axis in chicken controlled by different concentrations of BMP-4. Development. 1997;124:1975–84.
155. Tozer S, Duprez D. Tendon and ligament: development, repair and disease. Birth Defects Res C Embryo Today. 2005;75:226–36.
156. van den Akker E, Fromental-Ramain C, deGraaf W, LeMouellic H, Brulet P, Chambon P, Deschamps J. Axial skeletal patterning in mice lacking all paralogous group 8 Hox genes. Development. 2001;128:1911–21.
157. Veldhuizen AG, Wever DJ, Webb PJ. The aetiology of idiopathic scoliosis: biomechanical and neuromuscular factors. Eur Spine J. 2000;9:178–84.
158. Venters SJ, Thorsteinsdottir S, Duxson MJ. Early development of the myotome in the mouse. Dev Dyn. 1999;216:219–32.
159. Wagner J, Schmidt C, Nikowits W Jr, Christ B. Compartmentalization of the somite and myogenesis in chick embryos are influenced by wnt expression. Dev Biol. 2000;228:86–94.
160. Watanabe Y, Duprez D, Monsoro-Burq AH, Vincent C, Le Douarin NM. Two domains in vertebral development: antagonistic regulation by SHH and BMP4 proteins. Development. 1998;125:2631–9.
161. Wellik DM. *Hox* patterning of the vertebrate axial skeleton. Dev Dyn. 2007;236:2454–63.
162. Wellik DM, Capecchi MR. Hox10 and Hox11 genes are required to globally pattern the mammalian skeleton. Science. 2003;301:363–6.
163. Wood A, Thorogood P. Patterns of cell behavior underlying somitogenesis and notochord formation in intact vertebrate embryos. Dev Dyn. 1994;201:151–67.
164. Yang X, Arber S, William C, Li L, Tanabe Y, Jessell TM, Birchmeier C, Burden SJ. Patterning of muscle acetylcholine receptor gene expression in the absence of motor innervation. Neuron. 2001;30:399–410.

165. Yarom R, Robin GC. Studies on spinal and peripheral muscles from patients with scoliosis. Spine. 1979;4:12–21.
166. Yasuhiko Y, Haraguchi S, Kitajima S, Takahashi Y, Kanno J, Saga Y. Tbx6-mediated Notch signaling controls somite-specific Mesp2 expression. Proc Natl Acad Sci U S A. 2006;103:3651–6.
167. Yoon ST, Su Kim K, Li J, Soo Park J, Akamaru T, Elmer WA, et al. The effect of bone morphogenetic protein-2 on rat intervertebral disc cells in vitro. Spine. 2003;28:1773–80.
168. Zetterberg C, Aniansson A, Grimby G. Morphology of the paravertebral muscles in adolescent idiopathic scoliosis. Spine. 1983;8:457–62.
169. Zuk T. The role of spinal and abdominal muscles in the pathogenesis of scoliosis. J Bone Joint Surg Br. 1962;44:102–5.

Chapter 2
Environmental Factors and Axial Skeletal Dysmorphogenesis

Peter G. Alexander, Ricardo Londono, Thomas P. Lozito, and Rocky S. Tuan

Introduction

Axial skeletal development is part of the complex, inclusive process of axial or midline development. It involves the interaction of many tissues including the embryonic notochord, neural tube, somite compartments, intersomitic angiopotent cells, and neural crest cells. These tissues give rise to the axial skeleton, intervertebral discs, spinal cord, trunk musculature and dorsal dermis, intervertebral arteries, and spinal ganglia. Development of these tissues occurs in an interdependent and hierarchical manner over an extended period of time. These characteristics may make the axial skeleton disproportionately susceptible to environmental influence, accounting for the high incidence of axial skeletal defects among live and stillbirths. It may also account for the many manifestations axial skeletal defects observed.

Data show that the axial skeleton is one of several organ systems with a high frequency of abnormality, 1 in 1000 live births [27, 39, 50, 93, 147, 154], and a very low heritable component, estimated to be between 0.5 and 2%. Congenital axial skeletal defects may occur in isolation or as a component of more widespread syndromes or sequences [39, 50, 93, 154] (Table 2.1). It is estimated that the skeletal defect is accompanied by an intra-spinal neural defect in 40% of cases. In addition, approximately 50–60% of cases of congenital scoliosis suffer additional congenital defects in other organ systems including urogenital and cardiovascular systems (approximately 20% and 10–12%, respectively) and gastrointestinal and limb defects (2–5%). These combinations of congenital defects and their frequencies are reflective of the degree of concurrent development of the different organ systems.

P. G. Alexander · R. Londono · T. P. Lozito · R. S. Tuan (✉)
Center for Cellular and Molecular Engineering, Department of Orthopaedic Surgery, University of Pittsburgh School of Medicine, Pittsburgh, PA, USA
e-mail: rst13@pitt.edu

K. Kusumi, S. L. Dunwoodie (eds.), *The Genetics and Development of Scoliosis*, https://doi.org/10.1007/978-3-319-90149-7_2

Table 2.1 Genetic syndromes that are characterized by scoliosis

A representative list of recognized genetic syndromes that may include vertebral anomalies	
Syndrome	Features
Alagille syndrome (autosomal dominant)	Neonatal jaundice, cholestasis, peripheral pulmonic stenosis, occasional septal defects, and patent ductus arteriosus, accompanied by abnormal facies, ocular, *vertebral*, and nervous system abnormalities
Bertolotti syndrome	*Sacralization of the fifth lumbar vertebrae* with sciatica and *scoliosis*
Caudal dysgenesis syndrome	Failure *to form part or all of the coccygeal, sacral, and lumbar vertebrae* and corresponding spinal segments with malformation and dysfunction of the bowel and bladder
Cerebrocostomandibular syndrome (autosomal recessive)	Severe micrognathia, *severe costovertebral anomalies* including bell-shaped thorax, incompletely ossified, aberrant rib structure, abnormal rib connection to the vertebral body. Accompanied by palatal defects, glossoptosis, pre- and postnatal growth deficiencies, mental retardation
Coffin-Siris syndrome	Hypoplasia of the fifth fingers and toes associated with mental and growth retardation, coarse facies, mild microcephaly, hypotonia, lax joints, mild hirsutism and occasionally accompanied by cardiac, *vertebral*, and gastrointestinal abnormalities
Oculocerebral-hypopigmentation syndrome (autosomal recessive)	Oculocutaneous albinism, micropthalmus, opaque corneas, oligophrenia with spasticity, high-arched palate, gingival atrophy, *scoliosis*
Jarcho-Levin syndrome(spondylothoracic dysplasia) (autosomal recessive)	*Multiple vertebral defects*, short thorax, rib abnormalities, camptodactyly, syndactyly and accompanied by urogenital anomalies and respiratory dysfunction
Kabuki makeup syndrome	Mental retardation, dwarfism, *scoliosis*, cardiovascular abnormalities, and facies reminiscent of a Japanese Kabuki actor
King's syndrome (malignant hyperthermia)	Short stature, *kyphoscoliosis*, pectus carinatum, cryptorchidism, delayed motor development, progressive myopathy, structural cardiovascular defects
Klippel-Feil syndrome	Reduced number of cervical vertebrae, *cervical hemivertebrae*, low hairline, reduced neck mobility
Lenz's syndrome (X-linked)	Microphthalmia, anophthalmia, digital anomalies, narrow shoulders, double thumbs, *vertebral abnormalities*, dental, urogenital, and cardiovascular defects may occur
Multiple pterygium syndrome (autosomal recessive)	Pterygia of the neck, axillae, popliteal, antecubital, and intercrural areas, accompanied by hypertelorism, cleft palate, micrognathia, ptosis, short stature, and a wealth of skeletal anomalies including camptodactyly, syndactyly, equinovarus, rocker bottom feet, *vertebral fusions, and rib abnormalities*
Oculoauriculovertebral syndrome (Goldenhar syndrome)	Colobomas of the upper eyelids, bilateral accessory auricular appendages, *vertebral anomalies*, facial bossing, asymmetrical skull, low hairline, mandibular hypoplasia, low-set ears, and sometimes hemifacial microsomia

(continued)

Table 2.1 (continued)

A representative list of recognized genetic syndromes that may include vertebral anomalies	
Syndrome	Features
Rubenstein-Taybi syndrome	Mental and motor retardation, broad thumbs and big toes, short stature, high-arched palate, straight, beaked nose, various eye abnormalities, pulmonary stenosis, keloid formation at surgical scars, large foramen magnum, *vertebral and sternal abnormalities*
VATER-VACTERL sequence	*Vertebral anomalies*, anal atresia (cardiac abnormalities), tracheal fistula with esophageal atresia, renal defects (limb abnormalities)

While dramatic axial skeletal defects do occur in the context of syndromes and other anomalies, the majority of congenital spinal anomalies involve single structural defects of the spine and frequently few obvious coincident malformations or functional deficits [56, 93, 154], indicating that a time-dependent, tissue-specific insult may be involved. The complexity of axial skeletal development and the variety of axial skeletal defects suggest a variety of loci and mechanisms through which environmental factors may cause axial skeletal dysmorphogenesis.

Faced with the high social costs of resultant morbidity, it is critical to determine the possible impact any environmental factor may have on the embryo. Although many of the known human teratogens can produce axial skeletal defects, the etiology of over half of observed axial skeletal defects is unknown and is assumed to be multifactorial, a combination of genetic susceptibility and environmental insult [39, 93]. This fact highlights the need for investigating the role of environmental factors, alone or in combination, in the production of this particular class of defects. Such study requires the convergence of at least two broad fields of study. The first is developmental biology, to understand the details of normal development and identify new markers, loci, and perhaps possible mechanisms of teratogenesis. The second field is teratology, a discipline closely related to reproductive toxicology that involves assessing the impact of environmental factors on the new biological markers, loci, and mechanisms discovered and characterized in developmental biology.

Vertebral Dysmorphogenesis in Human Congenital Scoliosis

Clinically, congenital scoliosis is defined as a spinal curvature of over 10% caused by a structural vertebral defect [50, 56, 154]). The abnormal spinal curvature is further defined by its anterior-posterior location and the plane of curvature: coronal for scoliosis and sagittal for kyphosis. The characteristic feature of congenital axial skeletal defects is the malformation of vertebral bodies or processes evident at birth. Broadly, these vertebral defects are clinically classified as either failures in formation and morphogenesis represented by hemivertebrae, wedge vertebrae, open vertebral arches, bifid vertebrae, and vertebral agenesis or failures in segmentation

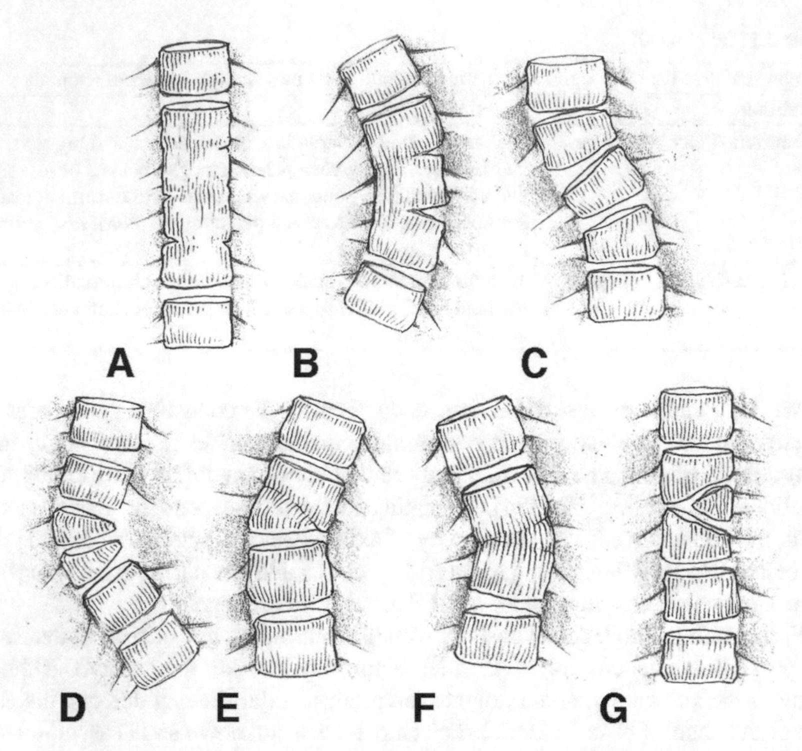

Fig. 2.1 Different forms of congenital scoliosis: block vertebrae (**a**); unilateral bar (**b**); wedge vertebrae (**c**); multiple hemivertebrae (**d**); single, semi-segmented vertebrae (**e**); non-segmented hemivertebrae (**f**); incarcerated hemivertebrae (**g**). Defects in segmentation form block vertebrae, vertebral bars. Defects in formation form hemivertebrae and wedge vertebrae. (Adapted from Parke [158])

represented by unilateral unsegmented bars or block vertebrae (Fig. 2.1) [50, 56, 53, 68, 93, 154]. Developmentally, these defects have their origin in somitogenesis, the initial manifestation of the vertebral column's metameric segmentation.

Normal Development of the Axial Skeleton

As discussed in other chapters of this volume, the axial skeleton is derived from the paraxial mesoderm, a primary germ layer, which undergoes the molecularly timed process of somitogenesis to produce blocks of tissue symmetrically arranged on either side of the midline neural tube and notochord (Table 2.2, Fig. 2.1) [38, 76, 186, 198]. The somite is a transient embryonic structure that plays an important role in the patterning of the axial skeleton (comprised of vertebral bodies, ribs, and intervertebral discs) and its associated tissues: the hypaxial and epaxial muscles of the spine, the dorsal dermis of the trunk, and the intervertebral arteries. The morphogenic description of somitogenesis can be conceptually divided into several phases: patterning, morphogenesis, differentiation, and growth and maturation [5, 38, 186, 204]

Table 2.2 Developmental timing of the axial skeleton in the human embryo

Developmental feature	Day of gestation	Other notable occurrences
Gastrulation	15	Neural plate formation
Notochord formation	17–19	Neural tube folding
First somite	19	Heart tube formation
Onset of neural tube fusion	22	Heart tube folding, optic and otic vesicle formation begins
Anterior neuropore closure	23–26	Embryonic circulation
Posterior neuropore closure	26–30	Forelimb bud
Sclerotomal segmentation	24–35	
Notochordal segmentation	28–30	
Last (30th) somite formed	32	Hindlimb bud, optic cup formed
All rib primordia evident	42–44	
Chondrification of centra	36–42	
Chondrification of ribs and laminae	40–44	
Chondrification complete/onset of ossification	56–60	

under tight temporospatial control [164]. These are helpful classifications when characterizing and studying birth defects and their causes.

Among the tissues of the spine, the axial skeleton and its composite tissues undergo multiple rounds of patterning, differentiation, and growth events including somitogenesis, resegmentation, and ossification, among other processes. Briefly, the axial skeleton is derived from the sclerotome, the ventromedial quadrant of each somite. Cells of sclerotome are initially part of the epithelial somite. Shortly after expressing the paired-box gene *Pax1* [16, 211], the cells de-epithelialize and relocate themselves to surround the notochord. These cells then begin expressing *Sox9*, a chondrocyte-specific transcription factor, and producing prodigious amounts of cartilage matrix to form the cartilage anlage of the vertebral body [81]. There is a distinct polarity to the somite as it matures [204] that is consequential during resegmentation, in which the posterior half of one somite merges with the anterior half of the posterior somite [37]. Together, these halves combine to form a vertebral body out of phase with the other tissue, characteristic of the vertebral motor unit [136, 214].

Development of the axial skeleton and the surrounding tissues occurs in an interdependent and hierarchical manner over an extended period of time. This may make the axial skeleton disproportionately susceptible to environmental influence, accounting for the high incidence of axial skeletal defects among live and stillbirths. It may also account for the many manifestations of axial skeletal defects observed [53, 70]. Understanding these processes (the normal development of the spine) and their effects upon the surrounding tissues is important in deciphering the etiology of various forms of congenital scoliosis and the mechanisms by which environmental agents may initiate abnormal development.

Experimental Axial Skeletal Teratology

Given that the majority of axial defects have no known genetic cause [39, 50, 56, 70, 93, 154], the assumption must be made that there is an environmental component. The principles that aid in the definition of an environmental teratological agent were defined and popularized in the wake of the "thalidomide experience" and with some modification remain applicable today [106, 179, 199, 208, 218]. In establishing the role of an environmental agent in inducing a congenital axial skeletal defect, we know that it must first affect the development and function of axial tissues and those that influence their differentiation including the notochord, neural tube, paraxial mesoderm, and overlying ectoderm. Second, the exposure must occur somewhere between the fourth and tenth week of human gestation, or organogenesis, during which time gastrulation, neurulation, and somitogenesis occur (Table 2.2) [145, 147]. Third, the target of the teratogen must play a necessary role in the affected developmental process (e.g., somitogenesis) by acting via a specific mechanism. Finally, we must observe a dose-response effect of the environmental agent on embryonic development in both frequency and degree that includes the graded manifestations of abnormal development: death, dysmorphogenesis, inhibition of growth or developmental delay, and functional deficit.

In the etiology of scoliosis, target organs may include the paraxial mesoderm and somites, the neural tube and notochord, and the overlying ectoderm. For example, the patterning of the somite boundaries and the subsequent boundaries of differentiation defined by integrated signaling pathways under the influence of the surrounding tissues figure prominently. Morphological processes that may be affected include somitogenesis, neurulation, and gastrulation, which involve cell migration, epithelialization, and laminar fusion, as well proliferation and apoptosis. Finally, the differentiation, growth, and maturation of the axial skeletal elements may also play an important role (Table 2.3).

Pathogenesis of Abnormal Axial Development

The identification of the structural defect in congenital scoliosis in the fetus or neonate remains an analysis conducted long after the initial pathogenic events inducing the malformation. Identifying and understanding the initial pathogenic event is a critical step in characterizing the mechanisms of teratogenesis, which can then lead to the development of appropriate interventions. Environmental insults to a developing organism occur at molecular or subcellular levels. While the list of possible environmental insults is very large, the insults may be translated into types of cellular responses that result in recognizable patterns of dysfunction of dysmorphogenesis among tissues and organs (Table 2.4).

Although teratogens are often discrete in nature (e.g., of known structure/composition and chemical characteristic), the determination of teratogenic mechanism is complicated. The main reasons for this are as follows. First, not all the possible targets of a

Table 2.3 Phases of somitogenesis in a stage 12 chick embryo and possible causal links between teratogen target tissues and hypothesized mature dysmorphogenesis

Stage 12 chick embryo [78]	Transverse section/time of teratogenic insult	Target tissue	Possible resultant dysmorphogenesis
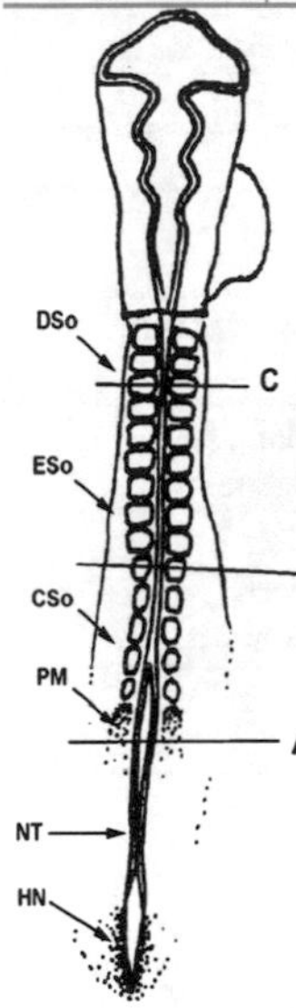	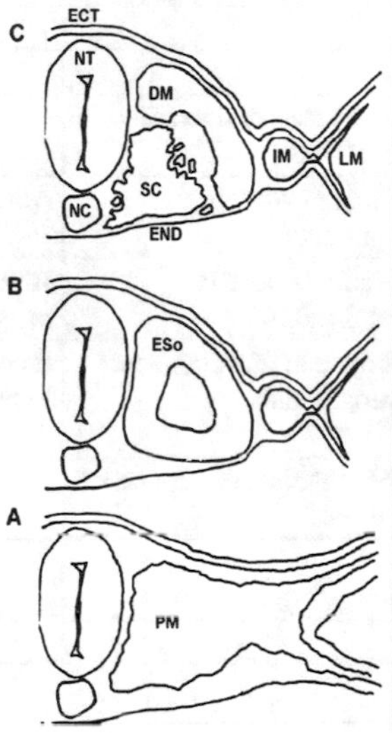 C	1. Notochord	Cleft vertebrae
		2. Ectoderm/ neural tube	Vertebral element agenesis
		3. Sclerotome	Vertebral disc anomalies Abnormal bone metabolism
	B	1. Notochord	Cleft vertebrae
		2.Ectoderm/ neural tube	Vertebral agenesis
		3. Somitic mesoderm	Hemivertebrae Block vertebrae
		4. Lateral plate mesoderm	Fused ribs Bifid ribs
	A	1. Chorda mesoderm/ notochord	Vertebral disc anomalies Caudal agenesis
		2. Paraxial mesoderm	Vertebral agenesis, hemi-block vertebrae
		3. Ectoderm	Block or hemivertebrae Bifid or fused ribs Vertebral agenesis

Labels: *DSo* differentiated somite, *ESo* epithelial somite, *CSo* condensed somite, *PM* paraxial mesoderm, *NT* neural tube, *HN* Hensen's node, *ECT* ectoderm, *END* endoderm, *NC* notochord, *DM* dermomyotome, *SC* sclerotome, *IM* intermediate mesoderm, *LM* lateral plate mesoderm

teratogen have been identified, since many potentially affected targets remain unknown, i.e., normal developmental mechanisms still need identification and characterization. Related to this is the fact that it is highly unlikely that most teratogens act upon a single molecule or even a cellular pathway. Multiple mechanistic pathways may combine to produce a single pathogenic mechanism contributing to the resultant congenital defect. Second, our ability to monitor the effect of the teratogen on the biochemistry of individual cellular targets is limited. Specifically, probes with sufficient sensitivity and specificity are unavailable for many processes and applications. Contributing to these issues is the fact that the amount of tissue available for study is usually very limited. Intertwined within these shortcomings is the difficulty of experimental interpretation, which varies with probe, detection methods, and the endpoint chosen.

Despite these complications, we can hypothesize several intracellular processes that may be targets of teratogens. The teratogen generates its effects on the embryo often through a mode of molecular mimicry co-opting or undermining normal cellular processes such that they are activated, inactivated, or diverted in a manner inconsistent with developmental timing. Such processes include mitotic interference (mutagenesis and carcinogenesis), epigenetic changes (methylation and acetylation state), altered membrane function (composition or porosity),

Table 2.4 Potential mechanisms, routes of pathogenesis, and ultimate morphogenetic outcome used by environmental teratogens in the induction of congenital malformation

Mechanisms →	Pathogenesis →	Pathway → final defect
Initial types of changes in developing cells or tissues after teratogenic insult:	*Manifested as one or more types of abnormal developmental processes:*	*Resulting common pathway to final morphogenic defect:*
Genetic mutation		
Chromosomal damage		
Epigenetic alteration	Increased or decreased cell death	
Mitotic interference	Failed cell-cell interactions	Altered patterning
Nucleic acid synthesis/balance	Reduced matrix biosynthesis	Abnormal morphogenesis
Altered enzymatic substrates, cofactors, etc.	Impeded morphogenic movements	Incomplete or imbalanced differentiation
Altered redox status	Mechanical disruption of tissues	
Disrupted membrane or cytoskeletal integrity		
Altered signal transduction		

Adapted from Wilson [218]

altered signal transduction, altered/inhibited energy metabolism, inhibition of waste (intermediary), metabolism, changed redox status, specific or general enzyme inhibition, and disturbances in nucleic acid synthesis, among many other possibilities. The cellular responses to these insults may be grouped into several common outcomes, including necrosis or apoptosis, reduced biosynthesis, failed cell/cell or cell/matrix interactions, impeded morphogenetic movement, and mechanical disruption of tissues. Ultimately, the final defect may be manifest via loss of cells or tissue or imbalances in growth and differentiation.

While specific mechanisms of many teratogenic insults remain largely unknown, the characterization of cellular responses has been more successful. One particularly well-characterized outcome is the correlation of tissue-specific patterns of cell death and impending malformation [79, 179, 199, 201, 226]. This correlation has highlighted several characteristics of teratogenic action including the principles that different cell populations are sensitive to teratogenic insults at different time points, different agents target different tissues, and many teratogens expand areas of normal, developmentally regulated cell death. The observed changes in normal cell death patterns indicate the target tissue and often play a role in the subsequent dysmorphogenesis [100, 162]; however, the apoptotic cells do not participate in subsequent tissue formation – thus the effect of the teratogen on the surviving cells is important and presumed to be related to the cause of cell death. Nonetheless, the increase in cell death serves as an early marker for the teratogenic action. As we learn more about development and toxicological responses on the molecular level, we will create more sensitive cell response markers that will allow greater resolution of the teratological action.

Overview of Agents and Conditions Associated with Axial Skeletal Teratogenesis

As stated above, axial skeletal malformations are often linked to exposure to teratogenic conditions. The following summarizes the types of teratogens and teratogenic conditions associated with spinal malformations [183]. Detailed descriptions of some of these factors will be presented in the following sections.

Recreational Teratogens

Recreational drugs such as alcohol and cocaine cigarettes are known to significantly reduce fetal and postnatal growth, significantly increase infant mortality, and cause congenital malformations of various types and severity.

Pharmaceutical Teratogens

Most embryonic organs and the central nervous system are extremely sensitive to the teratogenic effects of pharmaceuticals such as thalidomide, diethylstilbestrol, retinoic acid, valproic acid, warfarin, lithium, nicotinic acid, and many chemotherapeutics.

Industrial and Environmental Teratogens

Industrial processes required to fulfill the needs of growing populations worldwide release a substantial amount of waste products into the environment, with the toxicological and teratogenic effect of many species as-yet uncharacterized. Among the chemicals with known teratogenic effects are organic solvents, arsenic, cadmium and lead anesthetic gases, and organic mercury.

Agricultural Teratogens

Insecticides and herbicides are critical to providing nutrition to growing populations. Studies have determined that organochlorine insecticides such as DTT, parathion, and malathion may interfere with fertility and reproduction by mimicking estrogen-like compounds. Among herbicides, the by-product of Agent Orange 2,3,7,8-tetrachloro-dibenzo-*p*-dioxin (TCDD) is highly teratogenic causing cleft palate and congenital renal abnormalities.

Infectious Diseases

Microbial chemicals may act as teratogens. Microbes such as syphilis, cytomegalovirus, rubella, herpes, toxoplasma, and fifth disease affect 1–5% of all live births. These infections may cause a group of associated malformations known as the TORCH complex, as well as isolated structural defects and functional deficits.

Metabolic Conditions

Some metabolic disorders, most prominently diabetes and hyperthermia, also induce congenital malformations in the embryos. Diabetic pregnancy increases the frequency of a wide variety of congenital defects over background including cardiac defects, eye and ear defects, renal defects, and functional deficits in addition to a high rate of congenital scoliosis, increased embryonic death, and lifelong metabolic disorders.

Nongenetically Linked Conditions Characterized by Axial Skeletal Defects

Spina Bifida

One group of malformations in which axial skeletal defects are prevalent is in the spectrum of neural tube defects. Neural tube defects are frequently classified according to the severity of the neural lesion [147, 178, 184]. For example, spina bifida aperta (SBA) is associated with a failure of the neural arch to form over the underlying spinal column and is considered the mildest of neural tube defects. The most severe type of neural tube defect includes the type that is represented by spina bifida operta (SBO), which is manifested by an open spinal column and failure of most of the associated vertebral elements to form. It is thought that the failure in neural tube development is the primary cause of this malformation and that the malformation of the vertebral body is secondary, underscoring the interdependence of neural tube and somitic mesoderm morphogenesis in normal development [143]. The etiology of these abnormalities has a well-characterized environmental component. They may be caused by dietary deficiencies in folate [221], by fetal exposure to anticonvulsants such as phenytoin and valproic acid [59], and ethanol, among other causative agents [42]. The incidence of spina bifida can be reduced even in high-risk populations through proper nutrition and supplementation with antioxidants such as folic acid.

Table 2.5 Common features of different associations within the axial mesodermal dysplasia complex spectrum. Oculo-auriculo-vertebral (OAV), Polyoligodactyly-imperforate anus-vertebral anomalies syndrome (PIV), Pallister-Hall syndrome (PHS)

Malformation	VACTERL	VATER	OAV	PIV	PHS
Vertebral	X	X	X	X	
Imperforate anus	X	X		X	X
Craniofacial	X		X		
Tracheoesophageal fistula	X	X			
Renal abnormalities	X	X			X
Limb anomalies	X			X	X

The VATER Association

The VATER spectrum is a nonrandom association characterized by vertebral anomalies (V), anal atresia (At), tracheoesophageal (TE) fistula, and renal (R) anomalies [25, 39, 124]. This spectrum may also be associated with cardiovascular (C) anomalies and limb (L) anomalies (VACTERL). The incidence of VATER in diabetic mothers is 200× higher than in the general population which occurs at a rate of 16 per 100,000 births [39, 126, 160, 123]. Vertebral defects in this association can involve agenesis, hypoplasia, and hemivertebrae, often afflicting many contiguous vertebral units. As the acronym suggests, many associated tissues are affected.

The association of these different mesenchymal-derived tissues to the vertebral column and the timing of their development are critical to hypothesizing the origin and mechanism of the defect(s). Analyses of the frequency and co-occurrence of the features of VACTERL and other syndromes suggest that the anomalies can extend to various cranio-caudal levels suggesting a time dependency and critical period through a defect in a common mechanism of dysmorphogenesis [194, 197]. The VACTERL sequence can be conceptually included in a group of progressively severe spectrums of which it may be the most severe (Table 2.5). This broad spectrum of malformations has been coined the axial mesodermal dysplasia complex (AMDC) [197]. Some confounding features to any hypothesis are the broad range of defects sometimes involving tissues derived from all germ layers, its largely spontaneous occurrence, and low rate of subsequent inheritance.

There are two broad but related models currently employed to explain the etiology of AMDC suggesting that the collection of defects may arise from a single environmental insult at a time early in post-implantation development [23]. In the first theory, the embryo at the time of early gastrulation is comprised of a single morphogenetic field, the primary developmental field [126, 146]. At this time the embryo responds essentially as a single, homogeneous entity. The primary effect of the insult is to affect growth (proliferation) within the embryo, drastically affecting the existence and position of organizing centers and tissue morphogenesis throughout the embryo as this primary developmental field subdivides into secondary developmental fields that will give rise to the various organs and structures of the embryo including the axial skeleton [125]. If the insult occurs at this time, there is necessarily a broad range of structures affected (polytopic defects) often of mesenchymal origin but involving ectodermal and endodermal germ layers as well.

A second variant on the theory holds that the broad spectrum of defects reflects a common mechanistic cause in many tissues of a more heterogeneous entity comprised of multiple secondary developmental fields, such that different tissues of the embryo respond in specific manners to produce the wide spectrum of observed defects [24, 124]. The defect then is thought to lie more in mechanisms of patterning or morphogenesis as the insult or defect occurs slightly later in development. This latter variation on developmental field defects and the etiology of multiple congenital anomalies such as VATER appears to more easily explain the wide spectrum of cranio-caudal positions of the defects and the wide degree of severity observed in several multiple congenital defect associations by allowing for a longer critical period. Both theories have been characterized theoretically and statistically to the range of defects observed in infants born to diabetic mothers, one of the most frequently recognized "causative" factors of the VATER spectrum [126].

The high incidence of the VATER and other AMDC variants in diabetic mothers suggests an etiology that involves a fundamental metabolic imbalance in energy production or a dysfunction in a critical component of the embryonic stress response. Some investigators have suggested that the defects may arise from malformation or dysfunction of the notochord, which is critical to the establishment and maintenance of embryonic axes and the patterning and differentiation of many mesenchymal tissues [71]. It has been suggested that notochord mutants such as *Brachyury* or sonic hedgehog (*Shh*) knockouts could be used as models for VATER and AMDC [9]. We discuss the potential for energy metabolism dysregulation as the locus affected resulting in VATER in the context of diabetes-induced congenital scoliosis below.

Environmental Factors That Cause Axial Skeletal Dysmorphogenesis

Valproic Acid

Valproic acid (VPA) is an anti-epileptic drug which is associated with a 20-fold increased incidence of spina bifida, a neural tube defect, in children born to pregnant mothers undergoing VPA treatment [8, 103, 140, 149]. Experimentally, VPA has been shown to be teratogenic in mouse, rat, chick, hamster, rabbit, and rhesus monkeys [17, 18, 55, 82, 129, 210]. Skeletal abnormalities in these models were most commonly observed, involving vertebrae, ribs, digits, and craniofacial bones. These frequently occur in the context of other cardiovascular, urogenital, and neurological anomalies that together comprise the fetal valproate syndrome [94, 155]. The axial skeletal defects can include presacral vertebrae and cervical and thoracic ribs, indicating possible homeotic transformations. The defects may also include structural vertebral defects, indicating segmentation defects.

In general, the primary locus of teratogens causing spina bifida including VPA is believed to be the neural tube, resulting in failure of neural tube closure [207]. Subsequently, the neural arches are unable to fuse. However, vertebral defects such as block vertebrae and hemivertebrae sometimes coincident with a neural tube defect have also been observed following VPA exposure ([17]). More detailed studies have shown that important patterning genes, such as *PAX1* and paraxis, are downregulated by the administration of VPA in chick embryos [15, 17]. The malformations produced by VPA can be mimicked through the administration of antisense deoxynucleotides during somitogenesis [15, 17]. This type of data confirms that dysregulation of these genes can be teratogenic but does not indicate a specific mechanism for how this may occur.

The downregulation of these genes may be caused by decreased signaling or reduced or delayed differentiation caused by increased reactive oxygen species (ROS) production or altered nucleic acid metabolism [57, 141], as suggested by studies showing that folic acid administration can significantly reduce the incidence of experimentally induced VPA axial skeletal defects [59, 75, 94]. More recently, VPA has been shown to also inhibit histone deacetylase activity at therapeutic levels, and that this activity is correlated with axial skeletal defects and exencephaly [130]. In a comparison of the teratogenicity and changes in gene expression by VPA and Trichostatin A (TSA), many of the shared genetic effects were specific to skeletal and cardiac muscle, assigning a more specific mechanistic action of VPA to dysregulating epigenetic control which lead to altered gene expression.

Hypoxia

Congenital vertebral anomalies have been produced in newborn animals experimentally by transient hypoxia and transient exposure during the embryonic period [73, 89, 172, 215]. In these studies, many gross vertebral and associated skeletal defects have been induced, including hemivertebrae, vertebral fusions, fragmented vertebral bodies, bifid ribs, or junctions of two or more ribs. The nature and extent of skeletal malformations induced have been dependent upon the precise stage of somite formation at the time when maternal stress has been induced. Hypoxia is thought to affect the early embryo through the induction of increased ROS homotopically (where ROS are already prominent) [57] and later through altered vascularization [45, 72, 215]. Less well defined is the idea that hypoxia itself or its management is important in and of itself for morphogenic process or cell function during embryonic development [35, 185].

During early organogenesis as the embryonic circulation develops, the embryo is known to undergo a transition from anaerobic respiration to aerobic respiration [87, 88, 119, 131]. Recent studies have confirmed that oxygenation and the cellular response to oxygenation as interpreted through expression patterns of heat shock proteins (protective chaperones) [54, 134], antioxidant (superoxide dismutases) [65, 148, 216, 224, 225], and HIF1alpha expression [91, 92, 122, 133] vary between different tissues of the embryo over time. Some of these variations have been correlated to periods of teratogenic susceptibility [65, 152]. In this transition, mitochondrial respiration may be inefficient producing higher-than-usual amounts of ROS at a time when embryonic defenses against ROS damage are not well developed [108, 150]. This combination can lead to excess ROS-induced cell stress and cell death [29, 47, 51]. One hypothesis is that those tissues undergoing energetically demanding process such as morphogenesis are most susceptible to the oxygenation transition, a hypothesis furthered in diabetic embryopathy [137, 222]. The neural tube and somitic mesoderm have been shown to have a higher metabolic activity [118, 119, 131, 132, 167] than surrounding tissues during early organogenesis, the time of greatest susceptibility to environmentally induced axial skeletal defects. Recent research has begun to focus specifically on the interaction of the genetic variation and environmental stressors, in which the processes of neurulation [66] or somitogenesis [196] may be targeted by hypoxic exposure.

Carbon Monoxide

Early work studying the effects of hypoxia utilized carbon monoxide as a chemical hypoxic agent. Carbon monoxide (CO) is an odorless, colorless, non-irritating gas produced by the incomplete combustion of carbon containing materials. There have been no epidemiological studies of the direct effect of CO on human pregnancies [183]. However, there are several case reports and anecdotes suggesting that CO may be a teratogen in humans [116, 173]. Anecdotal accounts were given in Brander [173] and reported congenital malformations, such as microcephaly, micrognathia, and limb defects including hip dysplasia, tetraplegia, equinovarus, and limb reduction. Indirect epidemiological information can be obtained from the observations of pregnancy outcomes among women who smoke. Maternal smoking is associated with various adverse outcomes including low birth weight, decreases in successful births [62], and various behavioral defects which can be mimicked by CO alone in animal models [22].

There are a limited number of studies linking CO to congenital malformations. Early studies in chick, rabbit, and rat showed a causative relationship [14, 139]; however, later studies failed to confirm this connection [11]. More recent studies exploring threshold levels and critical periods related to CO-induced effects upon the embryo have documented CO-induced dysmorphogenesis [7, 13, 46, 58, 114]. CO exposures during early organogenesis, the critical period, resulted in vertebral anomalies, microphthalmia, and a phenotype similar to caudal dysgenesis syndrome. Such malformations have been reported with CO exposures administered during organogenesis in the context of other teratogens at sub-teratogenic levels [191, 192]. This may be a significant problem worldwide since acute carbon monoxide exposures may be higher and more frequent than often reported [62, 168, 171, 213].

CO does impair oxygen delivery to and into cells by binding hemoglobin, myoglobin, and other porphyrins; however, it may also function as a signaling molecule in the context of nitric oxide (NO) signaling [121]. When administered after the vascular system is developed, the axial defects caused by CO are attributed to vascular leakage and subsequent mechanical disruption of developing tissue [14]. However, during early organogenesis, axial defects involve the reduction of important segmentation genes including pax-1 and paraxis [7], resulting possibly in the impaired inductive interaction of the neural tube with the paraxial mesoderm with CO acting as a signaling molecule. Nitric oxide is known to regulate neurulation and other early embryonic processes [104], and CO can alter the production of NO in axial tissues [6]. The impaired interaction is likely due to a loss of cell function characterized or indicated by increased neural tube apoptosis and loss of neural tube-derived somite epithelialization signals.

Diabetes

Maternal diabetes is known to have many teratogenic effects [2, 64, 151]. Malformations including neural tube defects, caudal dysgenesis, vertebral defects, congenital heart defects, femoral hypoplasia, renal and craniofacial anomalies are described in infants of diabetic mothers. Caudal regression syndrome is a severe condition characterized by agenesis, regression, and/or disorganization of the posterior (sacral-lumbar) vertebrae and the malformation of the soft tissue at that level and below [23, 24, 126, 194]. It occurs 200 times more frequently in diabetic than in nondiabetic pregnancies [32, 61, 209]. Other major malformations of the midline are also much more frequent including VATER, OAV, and other major malformations. Together, these can be placed in a related and progressive spectrum of syndromes and nonrandom associations belonging to the axial mesodermal dysplasia complex (AMDC).

Mouse models utilizing "diabetic environments" or hyperglycemia report various anomalies encompassing the full spectrum of embryonic embryopathy [4, 68, 152, 194]. These models together reveal that hyperglycemia is sufficient to cause most of the defects observed in diabetic embryopathy including neural tube defects, axial skeletal defects, heart and craniofacial abnormalities, and lib and renal defects – although no individual model phenocopies the condition completely. At physiological levels of hyperglycemia or ketosis, the most consistent outcome is a failure of anterior and posterior neuropore closure [85, 153, 177, 178, 180]. Researchers have determined that the diabetic environment increases ROS production in these regions of the neural tube and in the primordia of the organs listed above including craniofacial region, otic and optic cups, Hensen's node, and the notochord, caused by the diabetic environment. Coincident with the high ROS are an increase in cell death and a decrease in Pax-3, a factor critical in neural tube closure [63, 115]. Application of folic acid and other antioxidants greatly reduced the incidence of ROS production [148], insipient cell death, and the reduction of *Pax3* expression.

The caudal agenesis/dysgenesis syndrome can be phenocopied by prolonged exposure to hyperglycemia, hyperketonemia, streptozotocin [61]. The myriad of defects in these severely affected animals indicate an early patterning event is disturbed [83, 194]. The notochord is laid down during gastrulation and is responsible for dorsoventral and mediolateral patterning as well as survival of the mesoderm during axis elongation. High levels of cell death in the notochord are observed in severely affected animals, suggesting the notochord function is likely to be compromised, and mutations in the T-box gene *Brachyury* [169] and disruption of *Shh* [99] and Wnt [161] signaling have been presented as possible models of caudal dysgenesis and other manifestations of ADMC.

The incidence and severity of malformations in diabetic pregnancies is correlated with poor glycemic control in the first trimester and can be reduced by instituting tight glycemic control prior to conception, and the evidence presented above of various antioxidants and insulin provides hope that a cocktail can be developed and delivered harmlessly to prevent the initiation of the diabetic embryopathic condition. While prevention of the condition appears at hand, the initial biochemical imbalance presents us with an interesting problem. The condition of hyperglycemia provides a "free" energy source that is readily available to the mitochondrion for ATP production [67], a condition opposite to hypoxia, in which ATP production in greatly decreased. A reasonable hypothesis incorporating these two conditions is that molecular regulation of any developmental process can be disturbed by abnormal maternal fuel metabolism, and the timing of specific episodes of poor glycemic control determines which organ systems are affected.

Retinoic Acid

Retinoic acid (RA) is an analogue of vitamin A commonly used to treat acne and other skin conditions. In humans, prenatal exposure results in a characteristic pattern of defects including abnormalities in the ears, mandibles, palates, aortic arch, and central nervous system. In animal models, many similar defects are observed [77]. During organogensis, higher doses of RA can induce axial skeletal defects and homeotic transformations [96, 98, 176] and even axial skeletal truncations [156]. Deficiency of retinoic acid also results in axial dysmorphogenesis [111], consistent with a role of retinoic acid in normal embryonic spine development.

As a naturally occurring chemical and signaling molecule, RA is involved in many aspects of embryonic patterning, including the patterning of the somites. The teratogenic effects of retinoic acid above are consistent with the in situ expression of RA receptors [43, 90, 120] and metabolic-transforming enzymes [30, 144, 170, 203] as well as the effect of knocking down these molecules in murine models. RA, its receptors, and CYP26 are expressed in the paraxial mesoderm and act as critical regulators in the coordination of the somitogenesis clock and HOX gene expression [49, 187]. At increasingly higher doses, it is hypothesized that RA interrupts tissue morphogenesis and neural crest migration and ultimately cell death in morphogenically critical tissues such as the neural tube, notochord, and paraxial mesoderm, resulting in a phenotype similar to caudal dysgenesis syndrome [90].

Hyperthermia

Exposure of the human fetus to high temperatures (e.g., 2 °C above normal), as in the case of high fever or prolonged hot tub usage, is associated with neural tube defects, heart defects, microphthalmia, and functional deficits [54, 74]. There is no epidemiological evidence suggesting heat shock causes axial skeletal defects in humans. In studying the mechanisms of heat shock teratogenesis in animal models, vertebral defects were observed in many species including mice, rats, and chicks [26, 80, 110, 135, 166]. The severity of these defects is correlated to the time and duration of exposure. Experimental studies in chick embryos revealed that at moderate levels and exposure times (42 °C for 20 min), one or two adjacent segments were fused into a single large somite. This effect was repeated every seven to eight somites separated by normal somitogenesis [166]. This result suggested a cell cycle-dependent mechanism to the defect and to somitogenesis itself, prompting the proposal of a clock and wave-front model for the patterning process of somitogenesis [165].

The response of the embryo to hyperthermia is very dependent upon the degree of temperature increase, its duration, and the stage at which the heat shock is experienced [74]. There is a steep threshold for embryonic survival and resorption, which suggests the general outcome of hyperthermia is embryonic resorption. At levels of hyperthermia inducing embryonic survival and malformation, tissue-specific cell death is observed. Investigators identified the induction of heat shock proteins (HSP) as a prominent feature of the embryonic response, including humans [36]. These molecular chaperones play important roles in regulating protein folding during normal cell function, but they also serve to protect cells from environmental insult. In the process, the HSP-bound proteins are not able to perform their function [28, 212]. During teratological doses of hyperthermia, the cell cycle is slowed, suggesting a mechanism of the vertebral anomalies observed. Recently the mechanism of somitogenesis was shown to involve the tightly controlled, cyclic expression of a variety of proteins many belonging to the Notch/Delta signaling system [190]. During heat shock, some of these proteins or their targets may be bound by HSP, and we can hypothesize that this would disrupt the somatogenic clock resulting in disrupted pattern and ultimately vertebral defects. An important feature of the protective heat shock response then and its relation to teratogenesis is that their activation and function may reduce or delay tissue development or morphogenic actions. In fact, many teratogenic insults induce HSP activity, and as such HSP activation may be an underlying commonality in teratogenic mechanisms along with ROS production and apoptosis.

Arsenic

Arsenic is a metal pollutant found naturally in groundwater and unnaturally in mine waste sites, industrial by-products, and agricultural runoff. It is toxic to humans and is known to cause birth defects including spina bifida, craniofacial defects, and developmental retardation and to decrease birth weight and increase incidences of fetal mortality, miscarriage, and stillbirth [48, 217, 219]. In experimental animal models, arsenic is teratogenic in mice, rats, and chicks [19, 33, 48, 84, 112, 163], with neural tube defects being common among all of them [188, 205]. Its toxicity is greatly dependent on its redox state: arsenate vs arsenite. The structure of arsenate can mimic that of phosphate groups, imparting arsenate with the ability to disrupt various cell processes including nucleic acid metabolism, lipid metabolism, and electron transport. Inefficiency electron transport can lead to high production of ROS which have documented cell destructive activities and teratogenic capacity [21, 86, 101, 113]. In addition, arsenate can be reduced to arsenite. The effects of arsenite on disruption of cell cycle and cytoskeletal structure have been attributed to its reaction to sulfhydryl groups [12, 107], which may account for its strong induction of the heat shock response [21, 69, 135]. In addition, arsenite can disrupt the citric acid cycle and electron transport via binding to thiol-group enzymatic active sites [48] and generating ROS and subsequent birth defects [109, 117, 182]. In fact, recent studies have identified chemicals that inhibit arsenic-induced cell and tissue responses in the embryo that also have antioxidant activity [3, 117, 195].

A disruption in the energy status of different tissues of the developing embryo is attributed to teratogenicity of arsenic causing similar malformations to those observed in hypoxic or hyperglycemic environment; however, arsenic has other distinctive effects on the embryo [48]. Arsenic and other metal compounds are very effective inducers of the heat shock response [21, 135], which may protect cells from molecular damage but induce birth defects in its own right via disruption of the cell cycle and other cyclic and time-dependent morphogenetic processes [220]. Recently, specific examples of arsenic effects on differentiation [105, 128], cell survival [193], and angiogenesis [44, 127] have been reported, disruption of which is known to cause congenital birth defects. In addition, cells surviving the initial arsenic insult may pass on genetic damage that contributes to subsequent carcinogenic transformation later in ontogeny [21, 95]. These multiple, interacting mechanisms may account for the wide range of malformations observed following acute arsenic exposure and increase susceptibility to other environmental insults at low, chronic exposure levels [138].

Ethanol

Ethanol, widely consumed as a recreational drug, has long been strongly associated with teratogenesis as fetal alcohol syndrome (FAS). FAS is present in up to 1 in 3 children of alcoholic mothers equal to 40,000 children per year [183, 206]. FAS manifestations include growth deficiency, central nervous system problems, characteristic facial features, and organ malformations. Features of FAS have been observed in animal models exposed to ethanol in utero or in vitro, including mice, rat, chick, and others [20, 34, 60, 157, 181, 183, 202, 223].

The mouse model has been a particularly effective model in elucidating the etiology of ethanol-induced birth defects. One mechanism of ethanol-induced teratogenesis is through ethanol impaired placental blood flow to the fetus by constricting blood vessels and inducing embryonic/fetal hypoxia and malnutrition [189]. Since ethanol rapidly crosses the placenta into the fetus, there are other direct embryonic and fetal targets of ethanol. Ethanol has also been shown to increase cell death in critical cell populations including anterior neural folds and neural crest cells [40, 52, 175, 201], which play critical role in the morphogenesis of the face. Neural crest cells are particularly vulnerable to ethanol, inducing delayed/altered migration and cell death [175].

Correlations have been made to increased ROS production within the neural crest population [102], mitochondrial dysfunction, and cell death in the etiology of ethanol and other teratogens [148]. The anterior neural tube and cranial neural crest have been the subject of intense scrutiny in the teratogenic mechanisms of ethanol; however, other tissues are also affected, including the eye, ear, heart, renal system, and axial skeleton [10, 31, 97, 159, 181, 200, 215]. With respect to the axial skeleton, investigators observed a misalignment or segmentation defect in ethanol-exposed embryos. Despite the substantial morphological difference with heat shock-treated embryos, the investigators suggested that the mechanism may be similar to heat shock [31], involving the induction of the stress response by increased ROS production.

Methanol

Another alcohol, methanol, is an alcohol encountered frequently during industrial processes, and the effects of inhaled methanol during pregnancy were compared to inhaled ethanol [142] in rats. At the highest doses, increases in external, visceral, and skeletal malformations were observed. Skeletal malformations were the most prevalent and included vertebral abnormalities the increased incidence of cervical ribs. Other skeletal abnormalities caused by methanol have been observed including holoprosencephaly, facial dysmorphogenesis, basicranial malformation, duplications of the atlas and axis and cervical vertebral abnormalities, and abnormal number of presacral vertebrae [41, 174]. Initial cellular responses appear similar to ethanol at the level of tissue-specific cell death [1]. In contrast to ethanol, many of the axial skeletal defects indicate homeotic shifts in segment identity.

Conclusion

Advances in cell and molecular biology with respect to normal development and somitogenesis and the pathogenesis and mechanisms of teratogenesis are occurring at a tremendous rate. This allows teratologists and developmental toxicologists the opportunity to revisit old problems with new tools. Despite the large number of cellular processes that may be disturbed by a teratogen, there are only a limited number of cellular and morphological outcomes. This has led investigators to strive for the identification of very defined critical periods and doses in a variety of model systems to aid in the identification of the initial targets of a teratogen and the true, hypothetically singular target molecule or process, as proposed by Wilson in 1956. Applying genomic and proteomic technologies to the problem of teratogenesis should begin reveal the full spectrum of cellular processes affected and elucidate links between variations in genotype and the effect of the environment on the phenotype that produce birth defects such as congenital scoliosis. The identification, at least in part, of this "holy grail" will aid in the development of new preventative treatments to a variety of teratogenic insults.

References

1. Abbott BD, Ebron-McCoy M, Andrews JE. Cell death in rat and mouse embryos exposed to methanol in whole embryo culture. Toxicology. 1995;97:159–71.
2. Aberg A, Westbom L, Källén B. Congenital malformations among infants whose mothers had gestational diabetes or preexisting diabetes. Early Hum Dev. 2001;61:85–95.
3. Ahmad M, Wadaa MA, Farooq M, Daghestani MH, Sami AS. Effectiveness of zinc in modulating perinatal effects of arsenic on the teratological effects in mice offspring. Biol Res. 2013;46:131–8. https://doi.org/10.4067/S0716-97602013000200003.
4. Akazawa S. Diabetic embryopathy: studies using a rat embryo culture system and an animal model. Congenit Anom (Kyoto). 2005;45:73–9.
5. Alexander PG, Boyce AT, Tuan RS. Skeletal development. In: Moody SA, editor. Principles of developmental genetics. New York: Elsevier Academic Press; 2007. p. 866–905.
6. Alexander PG, Chau L, Tuan RS. Role of nitric oxide in chick embryonic organogenesis and dysmorphogenesis. Birth Defects Res A Clin Mol Teratol. 2007;79:581–94. https://doi.org/10.1002/bdra.20386.
7. Alexander PG, Tuan RS. Carbon monoxide-induced axial skeletal dysmorphogenesis in the chick embryo. Birth Defects Res A Clin Mol Teratol. 2003;67:219–30. https://doi.org/10.1002/bdra.10041.
8. Alsdorf R, Wyszynski DF. Teratogenicity of sodium valproate. Expert Opin Drug Saf. 2005;4:345–53.
9. Arsic D, Qi BQ, Beasley SW. Hedgehog in the human: a possible explanation for the VATER association. J Paediatr Child Health. 2002;38:117–21.
10. Assadi FK, Zajac CS. Ultrastructural changes in the rat kidney following fetal exposure to ethanol. Alcohol. 1992;9:509–12.
11. Astrup P, Trolle D, Olsen HM, Kjeldsen K. Moderate hypoxia exposure and fetal development. Arch Environ Health. 1975;30:15–6.

12. Aung KH, Tsukahara S, Maekawa F, Nohara K, Nakamura K, Tanoue A. Role of environmental chemical insult in neuronal cell death and cytoskeleton damage. Biol Pharm Bull. 2015;38(8):1109–12. https://doi.org/10.1248/bpb.b14-00890.
13. Bailey LJ, Johnston MC, Billet J. Effects of carbon monoxide and hypoxia on cleft lip in A/J mice. Cleft Palate Craniofac J. 1995;32:14–9. https://doi.org/10.1597/1545-1569(1995)032<0014:EOCMAH>2.3.CO;2.
14. Baker FD, Tumasonis CF. Carbon monoxide and avian embryogenesis. Arch Environ Health. 1972;24:53–61.
15. Barnes GL, Alexander PG, Hsu CW, Mariani BD, Tuan RS. Cloning and characterization of chicken Paraxis: a regulator of paraxial mesoderm development and somite formation. Dev Biol. 1997;189:95–111. https://doi.org/10.1006/dbio.1997.8663.
16. Barnes GL, Hsu CW, Mariani BD, Tuan RS. Chicken Pax-1 gene: structure and expression during embryonic somite development. Differentiation. 1996;61:13–23. https://doi.org/10.1046/j.1432-0436.1996.6110013.x.
17. Barnes GL Jr, Mariani BD, Tuan RS. Valproic acid-induced somite teratogenesis in the chick embryo: relationship with Pax-1 gene expression. Teratology. 1996;54:93–102. https://doi.org/10.1002/(SICI)1096-9926(199606)54:2<93::AID-TERA5>3.0.CO;2-5.
18. Basu A, Wezeman FH. Developmental toxicity of valproic acid during embryonic chick vertebral chondrogenesis. Spine (Phila Pa 1976). 2000;25:2158–64.
19. Beaudoin AR. Teratogenicity of sodium arsenate in rats. Teratology. 1974;10:153–7.
20. Becker HC, Diaz-Granados JL, Randall CL. Teratogenic actions of ethanol in the mouse: a minireview. Pharmacol Biochem Behav. 1996;55:501–13.
21. Bernstam L, Nriagu J. Molecular aspects of arsenic stress. J Toxicol Environ Health B Crit Rev. 2000;3:293–322.
22. Bnait KS, Seller MJ. Ultrastructural changes in 9-day old mouse embryos following maternal tobacco smoke inhalation. Exp Toxicol Pathol. 1995;47:453–61. https://doi.org/10.1016/S0940-2993(11)80327-1.
23. Boer LL, Morava E, Klein WM, Schepens-Franke AN, Oostra RJ. Sirenomelia: a multisystemic polytopic field defect with ongoing controversies. Birth Defects Res. 2017;109:791–804. https://doi.org/10.1002/bdr2.1049.
24. Bohring A, Lewin SO, Reynolds JF, Voigtländer T, Rittinger O, Carey JC, et al. Polytopic anomalies with agenesis of the lower vertebral column. Am J Med Genet. 1999;87:99–114.
25. Botto LD, Khoury MJ, Mastroiacovo P, Castilla EE, Moore CA, Skjaerven R, et al. The spectrum of congenital anomalies of the VATER association: an international study. Am J Med Genet. 1997;71:8–15.
26. Breen JG, Claggett TW, Kimmel GL, Kimmel CA. Heat shock during rat embryo development *in vitro* results in decreased mitosis and abundant cell death. Reprod Toxicol. 1999;13:31–9.
27. Brent RL, Fawcett LB. Developmental toxicology, drugs, and fetal teratogenesis. In: Reece EA, Hobbins JC, editors. Clinical obstetrics: the fetus and mother. 3rd ed. Malden: Blackwell; 2007. p. 217–35.
28. Buckiová D, Kubínová L, Soukup A, Jelínek R, Brown NA. Hyperthermia in the chick embryo: HSP and possible mechanisms of developmental defects. Int J Dev Biol. 1998;42:737–40.
29. Burton GJ, Hempstock J, Jauniaux E. Oxygen, early embryonic metabolism and free radical-mediated embryopathies. Reprod Biomed Online. 2003;6:84–96.
30. Cammas L, Romand R, Fraulob V, Mura C, Dollé P. Expression of the murine retinol dehydrogenase 10 (Rdh10) gene correlates with many sites of retinoid signalling during embryogenesis and organ differentiation. Dev Dyn. 2007;236:2899–908. https://doi.org/10.1002/dvdy.21312.
31. Carvan MJ 3rd, Loucks E, Weber DN, Williams FE. Ethanol effects on the developing zebrafish: neurobehavior and skeletal morphogenesis. Neurotoxicol Teratol. 2004;26:757–68. https://doi.org/10.1016/j.ntt.2004.06.016.

32. Castori M, Silvestri E, Cappellacci S, Binni F, Sforzolini GS, Grammatico P. Sirenome lia and VACTERL association in the offspring of a woman with diabetes. Am J Med Genet A. 2010;152A(7):1803–7. https://doi.org/10.1002/ajmg.a.33460.
33. Chaineau E, Binet S, Pol D, Chatellier G, Meininger V. Embryotoxic effects of sodium arsenite and sodium arsenate on mouse embryos in culture. Teratology. 1990;41(1):105–12. https://doi.org/10.1002/tera.1420410111.
34. Chaudhuri JD. Alcohol and the developing fetus--a review. Med Sci Monit. 2000;6:1031–41.
35. Chen EY, Fujinaga M, Giaccia AJ. Hypoxic microenvironment within an embryo induces apoptosis and is essential for proper morphological development. Teratology. 1999;60:215–25. https://doi.org/10.1002/(SICI)1096-9926(199910)60:4<215::AID-TERA6>3.0.CO;2-2.
36. Child DF, Hudson PR, Hunter-Lavin C, Mukhergee S, China S, Williams CP, Williams JH. Birth defects and anti-heat shock protein 70 antibodies in early pregnancy. Cell Stress Chaperones. 2006;11:101–5.
37. Christ B, Huang R, Scaal M. Formation and differentiation of the avian sclerotome. Anat Embryol (Berl). 2004;208:333–50. https://doi.org/10.1007/s00429-004-0408-z.
38. Christ B, Huang R, Wilting J. The development of the avian vertebral column. Anat Embryol (Berl). 2000;202(3):179–94.
39. Cohen MM. The child with multiple birth defects. New York: Oxford University Press; 1997.
40. Coll TA, Tito LP, Sobarzo CM, Cebral E. Embryo developmental disruption during organogenesis produced by CF-1 murine periconceptional alcohol consumption. Birth Defects Res B Dev Reprod Toxicol. 2011;92:560–74. https://doi.org/10.1002/bdrb.20329.
41. Connelly LE, Rogers JM. Methanol causes posteriorization of cervical vertebrae in mice. Teratology. 1997;55:138–44. https://doi.org/10.1002/(SICI)1096-9926(199702)55:2<138::AID-TERA4>3.0.CO;2-#.
42. Copp AJ, Greene ND. Genetics and development of neural tube defects. J Pathol. 2010;220:217–30. https://doi.org/10.1002/path.2643.
43. Cui J, Michaille JJ, Jiang W, Zile MH. Retinoid receptors and vitamin A deficiency: differential patterns of transcription during early avian development and the rapid induction of RARs by retinoic acid. Dev Biol. 2003;260:496–511.
44. Cui Y, Han Z, Hu Y, Song G, Hao C, Xia H, Ma X. MicroRNA-181b and microRNA-9 mediate arsenic-induced angiogenesis via NRP1. J Cell Physiol. 2012;227:772–83. https://doi.org/10.1002/jcp.22789.
45. Danielson MK, Danielsson BR, Marchner H, Lundin M, Rundqvist E, Reiland S. Histopathological and hemodynamic studies supporting hypoxia and vascular disruption as explanation to phenytoin teratogenicity. Teratology. 1992;46(5):485–97. https://doi.org/10.1002/tera.1420460513.
46. Daughtrey WC, Newby-Schmidt MB, Norton S. Forebrain damage in chick embryos exposed to carbon monoxide. Teratology. 1983;28(1):83–9. https://doi.org/10.1002/tera.1420280111.
47. Dennery PA. Effects of oxidative stress on embryonic development. Birth Defects Res C Embryo Today. 2007;81:155–62. https://doi.org/10.1002/bdrc.20098.
48. DeSesso JM, Jacobson CF, Scialli AR, Farr CH, Holson JF. An assessment of the developmental toxicity of inorganic arsenic. Prod Toxicol. 1998;12:385–433.
49. Duester G. Retinoic acid regulation of the somitogenesis clock. Birth Defects Res C Embryo Today. 2007;81:84–92. https://doi.org/10.1002/bdrc.20092.
50. Dias MS. Normal and abnormal development of the spine. Neurosurg Clin N Am. 2007;18(3):415–29. https://doi.org/10.1016/j.nec.2007.05.003.
51. Dumollard R, Duchen M, Carroll J. The role of mitochondrial function in the oocyte and embryo. Curr Top Dev Biol. 2007;77:21–49. https://doi.org/10.1016/S0070-2153(06)77002-8.
52. Dunty WC Jr, Chen SY, Zucker RM, Dehart DB, Sulik KK. Selective vulnerability of embryonic cell populations to ethanol-induced apoptosis: implications for alcohol-related birth defects and neurodevelopmental disorder. Alcohol Clin Exp Res. 2001;25:1523–35.
53. Eckalbar WL, Fisher RE, Rawls A, Kusumi K. Scoliosis and segmentation defects of the vertebrae. Wiley Interdiscip Rev Dev Biol. 2012;1:401–23. https://doi.org/10.1002/wdev.34.

54. Edwards MJ, Walsh DA, Li Z. Hyperthermia, teratogenesis and the heat shock response in mammalian embryos in culture. Int J Dev Biol. 1997;41:345–58.
55. Ehlers K, Stürje H, Merker HJ, Nau H. Valproic acid-induced spina bifida: a mouse model. Teratology. 1992;45:145–54. https://doi.org/10.1002/tera.1420450208.
56. Erol B, Tracy MR, Dormans JP, Zackai EH, Maisenbacher MK, O'Brien ML, Turnpenny PD, Kusumi K. Congenital scoliosis and vertebral malformations: characterization of segmental defects for genetic analysis. J Pediatr Orthop. 2004; 24:674–82.
57. Fantel AG. Reactive oxygen species in developmental toxicity: review and hypothesis. Teratology. 1996;53:196–217. https://doi.org/10.1002/(SICI)1096-9926(199603)53:3<196::AID-TERA7>3.0.CO;2-2.
58. Farley FA, Hall J, Goldstein SA. Characteristics of congenital scoliosis in a mouse model. J Pediatr Orthop. 2006;26:341–6. https://doi.org/10.1097/01.bpo.0000203011.58529.d8.
59. Fathe K, Palacios A, Finnell RH. Brief report novel mechanism for valproate-induced teratogenicity. Birth Defects Res A Clin Mol Teratol. 2014;100:592–7. https://doi.org/10.1002/bdra.23277.
60. Fernandez K, Caul WF, Boyd JE, Henderson GI, Michaelis RC. Malformations and growth of rat fetuses exposed to brief periods of alcohol in utero. Teratog Carcinog Mutagen. 1983;3:457–60.
61. Ferrer-Vaquer A, Hadjantonakis AK. Birth defects associated with perturbations in preimplantation, gastrulation, and axis extension: from conjoined twinning to caudal dysgenesis. Wiley Interdiscip Rev Dev Biol. 2013;2(4):427–42. https://doi.org/10.1002/wdev.97.
62. Fichtner RR, Sullivan KM, Zyrkowski CL, Trowbridge FL. Racial/ethnic differences in smoking, other risk factors, and low birth weight among low-income pregnant women, 1978-1988. MMWR CDC Surveill Summ. 1990;39:13–21.
63. Fine EL, Horal M, Chang TI, Fortin G, Loeken MR. Evidence that elevated glucose causes altered gene expression, apoptosis, and neural tube defects in a mouse model of diabetic pregnancy. Diabetes. 1999;48:2454–62.
64. Finnell RH, Dansky LV. Parental epilepsy, anticonvulsant drugs, and reproductive outcome: epidemiologic and experimental findings spanning three decades; 1: animal studies. Reprod Toxicol. 1991;5(4):281–99.
65. Forsberg H, Borg LA, Cagliero E, Eriksson UJ. Altered levels of scavenging enzymes in embryos subjected to a diabetic environment. Free Radic Res. 1996;24:451–9.
66. Francesca LC, Claudia R, Molinario C, Annamaria M, Chiara F, Natalia C, et al. Variants in TNIP1, a regulator of the NF-kB pathway, found in two patients with neural tube defects. Childs Nerv Syst. 2016;32:1061–7. https://doi.org/10.1007/s00381-016-3087-1.
67. Galloway CA, Yoon Y. Mitochondrial dynamics in diabetic cardiomyopathy. Antioxid Redox Signal. 2015;22:1545–62. https://doi.org/10.1089/ars.2015.6293.
68. Geneviève D, de Pontual L, Amiel J, Sarnacki S, Lyonnet S. An overview of isolated and syndromic oesophageal atresia. Clin Genet. 2007;71:392–9. https://doi.org/10.1111/j.1399-0004.2007.00798.x.
69. German J, Louie E, Banerjee D. The heat-shock response *in vivo*: experimental induction during mammalian organogenesis. Teratog Carcinog Mutagen. 1986;6:555–62.
70. Giampietro PF, Dunwoodie SL, Kusumi K, Pourquié O, Tassy O, Offiah AC, et al. Progress in the understanding of the genetic etiology of vertebral segmentation disorders in humans. Ann N Y Acad Sci. 2009;1151:38–67. https://doi.org/10.1111/j.1749-6632.2008.03452.x.
71. Gilbert-Barness E, Debich-Spicer D, Cohen MM Jr, Opitz JM. Evidence for the "midline" hypothesis in associated defects of laterality formation and multiple midline anomalies. Am J Med Genet. 2001;101:382–7.
72. Grabowski CT. A quantitative study of the lethal and teratogenic effects of hypoxia on the three-day chick embryo. Am J Anat. 1961;109:25–35. https://doi.org/10.1002/aja.1001090104.
73. Grabowski CT, Paar JA. The teratogenic effects of graded doses of hypoxia on the chick embryo. Am J Anat. 1958;103(3):313–47. https://doi.org/10.1002/aja.1001030302.

74. Graham JM Jr, Edwards MJ, Edwards MJ. Teratogen update: gestational effects of maternal hyperthermia due to febrile illnesses and resultant patterns of defects in humans. Teratology. 1998;58:209–21. https://doi.org/10.1002/(SICI)1096-9926(199811)58:5<209::AID-TERA8>3.0.CO;2-Q.
75. Greene ND, Copp AJ. Mouse models of neural tube defects: investigating preventive mechanisms. Am J Med Genet C Semin Med Genet. 2005;135C:31–41. https://doi.org/10.1002/ajmg.c.30051.
76. Gridley T. The long and short of it: somite formation in mice. Dev Dyn. 2006;235:2330–6. https://doi.org/10.1002/dvdy.20850.
77. Gudas LJ. Retinoids and vertebrate development. J Biol Chem. 1994;269:15399–402.
78. Hamburger V, Hamilton HL. A series of normal stages in the development of the chick embryo. 1951. Dev Dyn. 1992;195:231–72. https://doi.org/10.1002/aja.1001950404.
79. Hansen JM, Harris C. Redox control of teratogenesis. Reprod Toxicol. 2013;35:165–79. https://doi.org/10.1016/j.reprotox.2012.09.004.
80. Harrouk WA, Wheeler KE, Kimmel GL, Hogan KA, Kimmel CA. Effects of hyperthermia and boric acid on skeletal development in rat embryos. Birth Defects Res B Dev Reprod Toxicol. 2005;74:268–76. https://doi.org/10.1002/bdrb.20047.
81. Healy C, Uwanogho D, Sharpe PT. Regulation and role of Sox9 in cartilage formation. Dev Dyn. 1999;215(1):69–78. https://doi.org/10.1002/(SICI)1097-0177(199905)215:1<69::AID-DVDY8>3.0.CO;2-N.
82. Hendrickx AG, Nau H, Binkerd P, Rowland JM, Rowland JR, Cukierski MJ, Cukierski MA. Valproic acid developmental toxicity and pharmacokinetics in the rhesus monkey: an interspecies comparison. Teratology. 1988;38:329–45. https://doi.org/10.1002/tera.1420380405.
83. Herion NJ, Salbaum JM, Kappen C. Traffic jam in the primitive streak: the role of defective mesoderm migration in birth defects. Birth Defects Res A Clin Mol Teratol. 2014;100:608–22. https://doi.org/10.1002/bdra.23283.
84. Hood RD, Bishop SL. Teratogenic effects of sodium arsenate in mice. Arch Environ Health. 1972;24:62–5.
85. Horton WE Jr, Sadler TW. Effects of maternal diabetes on early embryogenesis. Alterations in morphogenesis produced by the ketone body, B-hydroxybutyrate. Diabetes. 1983;32:610–6.
86. Hunter ES 3rd. Role of oxidative damage in arsenic-induced teratogenesis. Teratology. 2000;62:240. https://doi.org/10.1002/1096-9926(200010)62:4<240::AID-TERA14>3.0.CO;2-8.
87. Hunter ES 3rd, Sadler TW. Fuel-mediated teratogenesis: biochemical effects of hypoglycemia during neurulation in mouse embryos *in vitro*. Am J Phys. 1989;257:E269–76. https://doi.org/10.1152/ajpendo.1989.257.2.E269.
88. Hunter ES 3rd, Tugman JA. Inhibitors of glycolytic metabolism affect neurulation-staged mouse conceptuses *in vitro*. Teratology. 1995;52:317–23. https://doi.org/10.1002/tera.1420520602.
89. Ingalls TH, Philbrook FR. Monstrosities induced by hypoxia. N Engl J Med. 1958;259:558–64. https://doi.org/10.1056/NEJM195809182591202.
90. Iulianella A, Beckett B, Petkovich M, Lohnes D. A molecular basis for retinoic acid-induced axial truncation. Dev Biol. 1999;205:33–48. https://doi.org/10.1006/dbio.1998.9110.
91. Iyer NV, Kotch LE, Agani F, Leung SW, Laughner E, Wenger RH, et al. Cellular and developmental control of O2 homeostasis by hypoxia-inducible factor 1 alpha. Genes Dev. 1998;12:149–62.
92. Jain S, Maltepe E, Lu MM, Simon C, Bradfield CA. Expression of ARNT, ARNT2, HIF1 alpha, HIF2 alpha and Ah receptor mRNAs in the developing mouse. Mech Dev. 1998;73:117–23.
93. Jaskwhich D, Ali RM, Patel TC, Green DW. Congenital scoliosis. Curr Opin Pediatr. 2000;12:61–6.

94. Jentink J, Loane MA, Dolk H, Barisic I, Garne E, Morris JK, et al. Valproic acid monotherapy in pregnancy and major congenital malformations. N Engl J Med. 2010;362:2185–93. https://doi.org/10.1056/NEJMoa0907328.
95. Kaushal A, Zhang H, Karmaus WJJ, Everson TM, Marsit CJ, Karagas MR, et al. Genome-wide DNA methylation at birth in relation to in utero arsenic exposure and the associated health in later life. Environ Health. 2017;16:50. https://doi.org/10.1186/s12940-017-0262-0.
96. Kawanishi CY, Hartig P, Bobseine KL, Schmid J, Cardon M, Massenburg G, Chernoff N. Axial skeletal and Hox expression domain alterations induced by retinoic acid, valproic acid, and bromoxynil during murine development. J Biochem Mol Toxicol. 2003;17:346–56. https://doi.org/10.1002/jbt.10098.
97. Kennedy LA, Elliott MJ. Ocular changes in the mouse embryo following acute maternal ethanol intoxication. Int J Dev Neurosci. 1986;4:311–7.
98. Kessel M, Gruss P. Homeotic transformations of murine vertebrae and concomitant alteration of Hox codes induced by retinoic acid. Cell. 1991;67:89–104.
99. Kim PC, Mo R, Hui CC. Murine models of VACTERL syndrome: role of sonic hedgehog signaling pathway. J Pediatr Surg. 2001;36:381–4.
100. Kimura M, Ichimura S, Sasaki K, Masuya H, Suzuki T, Wakana S, et al. Endoplasmic reticulum stress-mediated apoptosis contributes to a skeletal dysplasia resembling platyspondylic lethal skeletal dysplasia, Torrance type, in a novel Col2a1 mutant mouse line. Biochem Biophys Res Commun. 2015;468:86–91. https://doi.org/10.1016/j.bbrc.2015.10.160.
101. Kitchin KT, Ahmad S. Oxidative stress as a possible mode of action for arsenic carcinogenesis. Toxicol Lett. 2003;137(1–2):3–13.
102. Kotch LE, Chen SY, Sulik KK. Ethanol-induced teratogenesis: free radical damage as a possible mechanism. Teratology. 1995;52:128–36. https://doi.org/10.1002/tera.1420520304.
103. Lammer EJ, Sever LE, Oakley GP Jr. Teratogen update: valproic acid. Teratology. 1987;35:465–73. https://doi.org/10.1002/tera.1420350319.
104. Lee QP, Juchau MR. Dysmorphogenic effects of nitric oxide (NO) and NO-synthase inhibition: studies with intra-amniotic injections of sodium nitroprusside and NG-monomethyl-L-arginine. Teratology. 1994;49:452–64. https://doi.org/10.1002/tera.1420490605.
105. Lencinas A, Broka DM, Konieczka JH, Klewer SE, Antin PB, Camenisch TD, Runyan RB. Arsenic exposure perturbs epithelial-mesenchymal cell transition and gene expression in a collagen gel assay. Toxicol Sci. 2010;116:273–85. https://doi.org/10.1093/toxsci/kfq086.
106. Leung MCK, Procter AC, Goldstone JV, Foox J, DeSalle R, Mattingly CJ, et al. Applying evolutionary genetics to developmental toxicology and risk assessment. Reprod Toxicol. 2017;69:174–86. https://doi.org/10.1016/j.reprotox.2017.03.003.
107. Levinson W, Oppermann H, Jackson J. Transition series metals and sulfhydryl reagents induce the synthesis of four proteins in eukaryotic cells. Biochim Biophys Acta. 1980;606:170–80.
108. Li R, Chase M, Jung SK, Smith PJ, Loeken MR. Hypoxic stress in diabetic pregnancy contributes to impaired embryo gene expression and defective development by inducing oxidative stress. Am J Physiol Endocrinol Metab. 2005;289:E591–9. https://doi.org/10.1152/ajpendo.00441.2004.
109. Li X, Ma Y, Li D, Gao X, Li P, Bai N, et al. Arsenic impairs embryo development via down-regulating Dvr1 expression in zebrafish. Toxicol Lett. 2012;212:161–8. https://doi.org/10.1016/j.toxlet.2012.05.011.
110. Li ZL, Shiota K. Stage-specific homeotic vertebral transformations in mouse fetuses induced by maternal hyperthermia during somitogenesis. Dev Dyn. 1999;216:336–48. https://doi.org/10.1002/(SICI)1097-0177(199912)216:4/5<336::AID-DVDY3>3.0.CO;2-5.
111. Li Z, Shen J, Wu WK, Wang X, Liang J, Qiu G, Liu J. Vitamin A deficiency induces congenital spinal deformities in rats. PLoS One. 2012;7:e46565. https://doi.org/10.1371/journal.pone.0046565.
112. Lindgren A, Danielsson BR, Dencker L, Vahter M. Embryotoxicity of arsenite and arsenate: distribution in pregnant mice and monkeys and effects on embryonic cells *in vitro*. Acta Pharmacol Toxicol (Copenh). 1984;54:311–20.

113. Ling M, Li Y, Xu Y, Pang Y, Shen L, Jiang R, et al. Regulation of miRNA-21 by reactive oxygen species-activated ERK/NF-κB in arsenite-induced cell transformation. Free Radic Biol Med. 2012;52:1508–18. https://doi.org/10.1016/j.freeradbiomed.2012.02.020.
114. Loder RT, Hernandez MJ, Lerner AL, Winebrener DJ, Goldstein SA, Hensinger RN, et al. The induction of congenital spinal deformities in mice by maternal carbon monoxide exposure. J Pediatr Orthop. 2000;20:662–6.
115. Loeken MR. Current perspectives on the causes of neural tube defects resulting from diabetic pregnancy. Am J Med Genet C Semin Med Genet. 2005;135C:77–87. https://doi.org/10.1002/ajmg.c.30056.
116. Longo LD. The biological effects of carbon monoxide on the pregnant woman, fetus, and newborn infant. Am J Obstet Gynecol. 1977;129:69–103.
117. Ma Y, Zhang C, Gao XB, Luo HY, Chen Y, Li HH, et al. Folic acid protects against arsenic-mediated embryo toxicity by up-regulating the expression of Dvr1. Sci Rep. 2015;5:16093. https://doi.org/10.1038/srep16093.
118. Mackler B, Grace R, Duncan HM. Studies of mitochondrial development during embryogenesis in the rat. Arch Biochem Biophys. 1971;144:603–10.
119. Mackler B, Grace R, Tippit DF, Lemire RJ, Shepard TH, Kelley VC. Studies of the development of congenital anomalies in rats. III. Effects of inhibition of mitochondrial energy systems on embryonic development. Teratology. 1975;12:291–6. https://doi.org/10.1002/tera.1420120311.
120. Maden M. Distribution of cellular retinoic acid-binding proteins I and II in the chick embryo and their relationship to teratogenesis. Teratology. 1994;50:294–301. https://doi.org/10.1002/tera.1420500404.
121. Maines MD. The heme oxygenase system: a regulator of second messenger gases. Annu Rev Pharmacol Toxicol. 1997;37:517–54. https://doi.org/10.1146/annurev.pharmtox.37.1.517.
122. Maltepe E, Schmidt JV, Baunoch D, Bradfield CA, Simon MC. Abnormal angiogenesis and responses to glucose and oxygen deprivation in mice lacking the protein ARNT. Nature. 1997;386:403–7. https://doi.org/10.1038/386403a0.
123. Martínez-Frías ML, Bermejo E, Rodríguez-Pinilla E, Prieto L, Frías JL. Epidemiological analysis of outcomes of pregnancy in gestational diabetic mothers. Am J Med Genet. 1998;78:140–5.
124. Martínez-Frías ML, Frías JL. Primary developmental field. III: clinical and epidemiological study of blastogenetic anomalies and their relationship to different MCA patterns. Am J Med Genet. 1997;70:11–5.
125. Martinez-Frias ML, Frias JL. VACTERL as primary, polytopic, developmental field defects. Am J Med Genet. 1999;83:13–6.
126. Martinez-Frias ML, Frias JL, Opitz JM. Errors of morphogenesis and developmental field theory. Am J Med Genet. 1998;76:291–6.
127. McCollum CW, Hans C, Shah S, Merchant FA, Gustafsson JÅ, Bondesson M. Embryonic exposure to sodium arsenite perturbs vascular development in zebrafish. Aquat Toxicol. 2014;152:152–63. https://doi.org/10.1016/j.aquatox.2014.04.006.
128. McCoy CR, Stadelman BS, Brumaghim JL, Liu JT, Bain LJ. Arsenic and its methylated metabolites inhibit the differentiation of neural plate border specifier cells. Chem Res Toxicol. 2015;28:1409–21. https://doi.org/10.1021/acs.chemrestox.5b00036.
129. Menegola E, Broccia ML, Nau H, Prati M, Ricolfi R, Giavini E. Teratogenic effects of sodium valproate in mice and rats at midgestation and at term. Teratog Carcinog Mutagen. 1996;16:97–108. https://doi.org/10.1002/(SICI)1520-6866(1996)16:2<97::AID-TCM4>3.0.CO;2-A.
130. Menegola E, Di Renzo F, Broccia ML, Giavini E. Inhibition of histone deacetylase as a new mechanism of teratogenesis. Birth Defects Res C Embryo Today. 2006;78:345–53. https://doi.org/10.1002/bdrc.20082.
131. Miki A, Fujimoto E, Ohsaki T, Mizoguti H. Effects of oxygen concentration on embryonic development in rats: a light and electron microscopic study using whole-embryo culture techniques. Anat Embryol (Berl). 1988;178:337–43.

132. Miki A, Mizoguchi A, Mizoguti H. Histochemical studies of enzymes of the energy metabolism in postimplantation rat embryos. Histochemistry. 1988;88:489–95.
133. Minet E, Michel G, Remacle J, Michiels C. Role of HIF-1 as a transcription factor involved in embryonic development, cancer progression and apoptosis (review). Int J Mol Med. 2000;5:253–9.
134. Mirkes PE. Molecular/cellular biology of the heat stress response and its role in agent-induced teratogenesis. Mutat Res. 1997;396:163–73.
135. Mirkes PE, Cornel L. A comparison of sodium arsenite- and hyperthermia-induced stress responses and abnormal development in cultured postimplantation rat embryos. Teratology. 1992;46(3):251–9. https://doi.org/10.1002/tera.1420460308.
136. Mittapalli VR, Huang R, Patel K, Christ B, Scaal M. Arthrotome: a specific joint forming compartment in the avian somite. Dev Dyn. 2005;234:48–53. https://doi.org/10.1002/dvdy.20502.
137. Moazzen H, Lu X, Liu M, Feng Q. Pregestational diabetes induces fetal coronary artery malformation via reactive oxygen species signaling. Diabetes. 2015;64:1431–43. https://doi.org/10.2337/db14-0190.
138. Moreira A, Freitas R, Figueira E, Volpi Ghirardini A, Soares AMVM, Radaelli M, et al. Combined effects of arsenic, salinity and temperature on Crassostrea gigas embryotoxicity. Ecotoxicol Environ Saf. 2018;147:251–9. https://doi.org/10.1016/j.ecoenv.2017.08.043.
139. Murray FJ, Schwetz BA, Crawford AA, Henck JW, Quast JF, Staples RE. Embryotoxicity of inhaled sulfur dioxide and carbon monoxide in mice and rabbits. J Environ Sci Health C. 1979;13(3):233–50.
140. Nau H. Valproic acid-induced neural tube defects. Ciba Found Symp. 1994;181:144–52.
141. Nau H, Hauck RS, Ehlers K. Valproic acid-induced neural tube defects in mouse and human: aspects of chirality, alternative drug development, pharmacokinetics and possible mechanisms. Pharmacol Toxicol. 1991;69:310–21.
142. Nelson BK, Brightwell WS, MacKenzie DR, Khan A, Burg JR, Weigel WW, Goad PT. Teratological assessment of methanol and ethanol at high inhalation levels in rats. Fundam Appl Toxicol. 1985;5:727–36.
143. Nikolopoulou E, Galea GL, Rolo A, Greene ND, Copp AJ. Neural tube closure: cellular, molecular and biomechanical mechanisms. Development. 2017;144:552–66. https://doi.org/10.1242/dev.145904.
144. Niederreither K, Fraulob V, Garnier JM, Chambon P, Dollé P. Differential expression of retinoic acid-synthesizing (RALDH) enzymes during fetal development and organ differentiation in the mouse. Mech Dev. 2002;110:165–71.
145. Nishimura H, Tanimura T, Semba R, Uwabe C. Normal development of early human embryos: observation of 90 specimens at Carnegie stages 7 to 13. Teratology. 1974;10:1–5. https://doi.org/10.1002/tera.1420100102.
146. Opitz JM, Zanni G, Reynolds JF Jr, Gilbert-Barness E. Defects of blastogenesis. Am J Med Genet. 2002;11:269–86.
147. O'Rahilly RR, Mueller F. Human embryology and teratology. 3rd ed. New York: Wiley-Liss Publishers; 1996.
148. Ornoy A. Embryonic oxidative stress as a mechanism of teratogenesis with special emphasis on diabetic embryopathy. Reprod Toxicol. 2007;24:31–41. https://doi.org/10.1016/j.reprotox.2007.04.004.
149. Ornoy A. Valproic acid in pregnancy: how much are we endangering the embryo and fetus? Reprod Toxicol. 2009;28:1–10. https://doi.org/10.1016/j.reprotox.2009.02.014.
150. Ornoy A, Rand SB, Bischitz N. Hyperglycemia and hypoxia are interrelated in their teratogenic mechanism: studies on cultured rat embryos. Birth Defects Res B Dev Reprod Toxicol. 2010;89:106–15. https://doi.org/10.1002/bdrb.20230.
151. Ornoy A, Reece EA, Pavlinkova G, Kappen C, Miller RK. Effect of maternal diabetes on the embryo, fetus, and children: congenital anomalies, genetic and epigenetic changes and

developmental outcomes. Birth Defects Res C Embryo Today. 2015;105:53–72. https://doi.org/10.1002/bdrc.21090.

152. Ornoy A, Zaken V, Kohen R. Role of reactive oxygen species (ROS) in the diabetes-induced anomalies in rat embryos *in vitro*: reduction in antioxidant enzymes and low-molecular-weight antioxidants (LMWA) may be the causative factor for increased anomalies. Teratology. 1999;60:376–86. https://doi.org/10.1002/(SICI)1096-9926(199912)60:6<376::AID-TERA10>3.0.CO;2-Q.
153. Ornoy A, Zusman I, Cohen AM, Shafrir E. Effects of sera from Cohen, genetically determined diabetic rats, streptozotocin diabetic rats and sucrose fed rats on *in vitro* development of early somite rat embryos. Diabetes Res. 1986;3:43–51.
154. Oskouian RJ Jr, Sansur CA, Shaffrey CI. Congenital abnormalities of the thoracic and lumbar spine. Neurosurg Clin N Am. 2007;18:479–98. https://doi.org/10.1016/j.nec.2007.04.004.
155. Ozkan H, Cetinkaya M, Köksal N, Yapici S. Severe fetal valproate syndrome: combination of complex cardiac defect, multicystic dysplastic kidney, and trigonocephaly. J Matern Fetal Neonatal Med. 2011;24:521–4. https://doi.org/10.3109/14767058.2010.501120.
156. Padmanabhan R. Retinoic acid-induced caudal regression syndrome in the mouse fetus. Reprod Toxicol. 1998;12:139–51.
157. Padmanabhan R, Muawad WM. Exencephaly and axial skeletal dysmorphogenesis induced by acute doses of ethanol in mouse fetuses. Drug Alcohol Depend. 1985;16:215–27.
158. Parke WW. Development of the spine. In: Herkowitz HN, Garfin SR, Balderston RA, Eismont FJ, Bell GR, Weisel SW, editors. Rothman-Simeone: The Spine. 4th ed. Philadelphia, PA: W. B. Saunders Company; 1999. p. 3–29.
159. Parnell SE, Dehart DB, Wills TA, Chen SY, Hodge CW, Besheer J, et al. Maternal oral intake mouse model for fetal alcohol spectrum disorders: ocular defects as a measure of effect. Alcohol Clin Exp Res. 2006;30:1791–8. https://doi.org/10.1111/j.1530-0277.2006.00212.x.
160. Pauli RM. Lower mesodermal defects: a common cause of fetal and early neonatal death. Am J Med Genet. 1994;50:154–72.
161. Pavlinkova G, Salbaum JM, Kappen C. Wnt signaling in caudal dysgenesis and diabetic embryopathy. Birth Defects Res A Clin Mol Teratol. 2008;82:710–9. https://doi.org/10.1002/bdra.20495.
162. Pennimpede T, Proske J, König A, Vidigal JA, Morkel M, Bramsen JB, et al. *In vivo* knockdown of Brachyury results in skeletal defects and urorectal malformations resembling caudal regression syndrome. Dev Biol. 2012;372:55–67. https://doi.org/10.1016/j.ydbio.2012.09.003.
163. Peterková R, Puzanová L. Effect of trivalent and pentavalent arsenic on early developmental stages of the chick embryo. Folia Morphol (Praha). 1976;24:5–13.
164. Pourquié O. Vertebrate segmentation: from cyclic gene networks to scoliosis. Cell. 2011;145:650–63.
165. Primmett DR, Norris WE, Carlson GJ, Keynes RJ, Stern CD. Periodic segmental anomalies induced by heat shock in the chick embryo are associated with the cell cycle. Development. 1989;105:119–30.
166. Primmett DR, Stern CD, Keynes RJ. Heat shock causes repeated segmental anomalies in the chick embryo. Development. 1988;104:331–9.
167. Raddatz E, Kucera P. Mapping of the oxygen consumption in the gastrulating chick embryo. Respir Physiol. 1983;51:153–66.
168. Ralston JD, Hampson NB. Incidence of severe unintentional carbon monoxide poisoning differs across racial/ethnic categories. Public Health Rep. 2000;115:46–51.
169. Rashbass P, Wilson V, Rosen B, Beddington RS. Alterations in gene expression during mesoderm formation and axial patterning in Brachyury (T) embryos. Int J Dev Biol. 1994;38:35–44.
170. Reijntjes S, Gale E, Maden M. Generating gradients of retinoic acid in the chick embryo: Cyp26C1 expression and a comparative analysis of the Cyp26 enzymes. Dev Dyn. 2004;230:509–17. https://doi.org/10.1002/dvdy.20025.

171. Ritz B, Wilhelm M. Ambient air pollution and adverse birth outcomes: methodologic issues in an emerging field. Basic Clin Pharmacol Toxicol. 2008;102:182–90. https://doi.org/10.1111/j.1742-7843.2007.00161.x.
172. Rivard CH. Effects of hypoxia on the embryogenesis of congenital vertebral malformations in the mouse. Clin Orthop Relat Res. 1986;(208):126–30.
173. Robkin MA. Carbon monoxide and the embryo. Int J Dev Biol. 1997;41:283–9.
174. Rogers JM, Brannen KC, Barbee BD, Zucker RM, Degitz SJ. Methanol exposure during gastrulation causes holoprosencephaly, facial dysgenesis, and cervical vertebral malformations in C57BL/6J mice. Birth Defects Res B Dev Reprod Toxicol. 2004;71:80–8. https://doi.org/10.1002/bdrb.20003.
175. Rovasio RA, Battiato NL. Role of early migratory neural crest cells in developmental anomalies induced by ethanol. Int J Dev Biol. 1995;39(2):421–2.
176. Rubin WW, LaMantia AS. Age-dependent retinoic acid regulation of gene expression distinguishes the cervical, thoracic, lumbar, and sacral spinal cord regions during development. Dev Neurosci. 1999;21:113–25. https://doi.org/10.1159/000017373.
177. Sadler TW, Horton WE Jr. Effects of maternal diabetes on early embryogenesis. The role of insulin and insulin therapy. Diabetes. 1983;32:1070–4.
178. Sadler TW, Hunter ES 3rd, Balkan W, Horton WE Jr. Effects of maternal diabetes on embryogenesis. Am J Perinatol. 1988;5(4):319–26. https://doi.org/10.1055/s-2007-999717.
179. Sadler TW, Hunter ES 3rd. Principles of abnormal development. In: Kimmel CA, Buelke-Sam J, editors. Developmental toxicology. 2nd ed. New York: Raven Press; 1994. p. 53–63.
180. Sadler TW, Hunter ES 3rd, Wynn RE, Phillips LS. Evidence for multifactorial origin of diabetes-induced embryopathies. Diabetes. 1989;38:70–4.
181. Sanders EJ, Cheung E. Ethanol treatment induces a delayed segmentation anomaly in the chick embryo. Teratology. 1990;41:289–97. https://doi.org/10.1002/tera.1420410306.
182. Sannadi S, Kadeyala PK, Gottipolu RR. Reversal effect of monoisoamyl dimercaptosuccinic acid (MiADMSA) for arsenic and lead induced perturbations in apoptosis and antioxidant enzymes in developing rat brain. Int J Dev Neurosci. 2013;31:586–97. https://doi.org/10.1016/j.ijdevneu.2013.07.003.
183. Schardein JL. Chemically induced birth defects. 3rd ed. New York: Marcel Dekker; 2000.
184. Schwartz ES, Rossi A. Congenital spine anomalies: the closed spinal dysraphisms. Pediatr Radiol. 2015;45(Suppl 3):S413–9. https://doi.org/10.1007/s00247-015-3425-6.
185. Semenza GL. Perspectives on oxygen sensing. Cell. 1999;98:281–4.
186. Senthinathan B, Sousa C, Tannahill D, Keynes R. The generation of vertebral segmental patterning in the chick. J Anat. 2012;220:591–602. https://doi.org/10.1111/j.1469-7580.2012.01497.x.
187. Sewell W, Kusumi K. Genetic analysis of molecular oscillators in mammalian somitogenesis: clues for studies of human vertebral disorders. Birth Defects Res C Embryo Today. 2007;81:111–20. https://doi.org/10.1002/bdrc.20091.
188. Shalat SL, Walker DB, Finnell RH. Role of arsenic as a reproductive toxin with particular attention to neural tube defects. J Toxicol Environ Health. 1996;48(3):253–72. https://doi.org/10.1080/009841096161320.
189. Shibley IA Jr, McIntyre TA, Pennington SN. Experimental models used to measure direct and indirect ethanol teratogenicity. Alcohol Alcohol. 1999;34(2):125–40.
190. Shifley ET, Cole SE. The vertebrate segmentation clock and its role in skeletal birth defects. Birth Defects Res C Embryo Today. 2007;81:121–33. https://doi.org/10.1002/bdrc.20090.
191. Singh J. Interaction of maternal protein and carbon monoxide on pup mortality in mice: implications for global infant mortality. Birth Defects Res B Dev Reprod Toxicol. 2006;77(3):216–26. https://doi.org/10.1002/bdrb.20077.
192. Singh J, Aggison L Jr, Moore-Cheatum L. Teratogenicity and developmental toxicity of carbon monoxide in protein-deficient mice. Teratology. 1993;48:149–59. https://doi.org/10.1002/tera.1420480209.
193. Singh S, Greene RM, Pisano MM. Arsenate-induced apoptosis in murine embryonic maxillary mesenchymal cells via mitochondrial-mediated oxidative injury. Birth Defects Res A Clin Mol Teratol. 2010;88:25–34. https://doi.org/10.1002/bdra.20623.

194. Solomon BD, Bear KA, Kimonis V, de Klein A, Scott DA, Shaw-Smith C, et al. Clinical geneticists' views of VACTERL/VATER association. Am J Med Genet A. 2012;158A:3087–100. https://doi.org/10.1002/ajmg.a.35638.
195. Song G, Cui Y, Han ZJ, Xia HF, Ma X. Effects of choline on sodium arsenite-induced neural tube defects in chick embryos. Food Chem Toxicol. 2012;50:4364–74. https://doi.org/10.1016/j.fct.2012.08.023.
196. Sparrow DB, Chapman G, Smith AJ, Mattar MZ, Major JA, O'Reilly VC, et al. A mechanism for gene-environment interaction in the etiology of congenital scoliosis. Cell. 2012;149:295–306. https://doi.org/10.1016/j.cell.2012.02.054.
197. Stewart FJ, Nevin NC, Brown S. Axial mesodermal dysplasia spectrum. Am J Med Genet. 1993;45:426–9. https://doi.org/10.1002/ajmg.1320450405.
198. Stockdale FE, Nikovits W Jr, Christ B. Molecular and cellular biology of avian somite development. Dev Dyn. 2000;219(3):304–21. https://doi.org/10.1002/1097-0177(2000)9999:9999<::AID-DVDY1057>3.0.CO;2-5.
199. Sulik KK. Pathogenesis of abnormal development. In: Hood RD, editor. Handbook of developmental toxicology. New York: CRC Press; 1997. p. 43–60.
200. Sulik KK. Genesis of alcohol-induced craniofacial dysmorphism. Exp Biol Med (Maywood). 2005;230:366–75.
201. Sulik KK, Cook CS, Webster WS. Teratogens and craniofacial malformations: relationships to cell death. Development. 1988;103(Suppl):213–31.
202. Sulik KK, Johnston MC, Webb MA. Fetal alcohol syndrome: embryogenesis in a mouse model. Science. 1981;214:936–8.
203. Swindell EC, Thaller C, Sockanathan S, Petkovich M, Jessell TM, Eichele G. Complementary domains of retinoic acid production and degradation in the early chick embryo. Dev Biol. 1999;216:282–96. https://doi.org/10.1006/dbio.1999.9487.
204. Tam PP, Trainor PA. Specification and segmentation of the paraxial mesoderm. Anat Embryol (Berl). 1994;189:275–305.
205. Takeuchi IK. Embryotoxicity of arsenic acid: light and electron microscopy of its effect on neurulation-stage rat embryo. J Toxicol Sci. 1979;4:405–16.
206. Thackray H, Tifft C. Fetal alcohol syndrome. Pediatr Rev. 2001;22:47–55.
207. Turner S, Sucheston ME, De Philip RM, Paulson RB. Teratogenic effects on the neuroepithelium of the CD-1 mouse embryo exposed in utero to sodium valproate. Teratology. 1990;41(4):421–42. https://doi.org/10.1002/tera.1420410408.
208. Ujházy E, Mach M, Navarová J, Brucknerová I, Dubovický M. Teratology - past, present and future. Interdiscip Toxicol. 2012;5:163–8. https://doi.org/10.2478/v10102-012-0027-0.
209. Versiani BR, Gilbert-Barness E, Giuliani LR, Peres LC, Pina-Neto JM. Caudal dysplasia sequence: severe phenotype presenting in offspring of patients with gestational and pregestational diabetes. Clin Dysmorphol. 2004;13:1–5.
210. Vorhees CV. Teratogenicity and developmental toxicity of valproic acid in rats. Teratology. 1987;35:195–202. https://doi.org/10.1002/tera.1420350205.
211. Wallin J, Wilting J, Koseki H, Fritsch R, Christ B, Balling R. The role of Pax-1 in axial skeleton development. Development. 1994;120(5):1109–21.
212. Walsh D, Grantham J, Zhu XO, Wei Lin J, van Oosterum M, Taylor R, Edwards M. The role of heat shock proteins in mammalian differentiation and development. Environ Med. 1999;43:79–87.
213. Wang L, Pinkerton KE. Air pollutant effects on fetal and early postnatal development. Birth Defects Res C Embryo Today. 2007;81(3):144–54. https://doi.org/10.1002/bdrc.20097.
214. Ward L, Evans SE, Stern CD. A resegmentation-shift model for vertebral patterning. J Anat. 2017;230:290–6. https://doi.org/10.1111/joa.12540.
215. Webster WS, Abela D. The effect of hypoxia in development. Birth Defects Res C Embryo Today. 2007;81:215–28. https://doi.org/10.1002/bdrc.20102.
216. Wells PG, Winn LM. Biochemical toxicology of chemical teratogenesis. Crit Rev Biochem Mol Biol. 1996;31:1–40. https://doi.org/10.3109/10409239609110574.

217. Willhite CC, Ferm VH. Prenatal and developmental toxicology of arsenicals. Adv Exp Med Biol. 1984;177:205–28.
218. Wilson JG. Current status of teratology: general principles and mechanisms derived from animal studies. In: Wilson JG, Fraser CF, editors. Handbook of teratology. New York: Plenum Press; 1997. p. 47.
219. Winterbottom EF, Fei DL, Koestler DC, Giambelli C, Wika E, Capobianco AJ, et al. GLI3 links environmental arsenic exposure and human fetal growth. EBioMedicine. 2015;2:536–43. https://doi.org/10.1016/j.ebiom.2015.04.019.
220. Wlodarczyk BJ, Bennett GD, Calvin JA, Finnell RH. Arsenic-induced neural tube defects in mice: alterations in cell cycle gene expression. Reprod Toxicol. 1996;10:447–54.
221. Yamaguchi Y, Miyazawa H, Miura M. Neural tube closure and embryonic metabolism. Congenit Anom (Kyoto). 2017;57:134–7. https://doi.org/10.1111/cga.12219.
222. Yang P, Reece EA, Wang F, Gabbay-Benziv R. Decoding the oxidative stress hypothesis in diabetic embryopathy through proapoptotic kinase signaling. Am J Obstet Gynecol. 2015;212:569–79. https://doi.org/10.1016/j.ajog.2014.11.036.
223. Yelin R, Kot H, Yelin D, Fainsod A. Early molecular effects of ethanol during vertebrate embryogenesis. Differentiation. 2007;75:393–403. https://doi.org/10.1111/j.1432-0436.2006.00147.x.
224. Yon JM, Baek IJ, Lee SR, Jin Y, Kim MR, Nahm SS, et al. The spatio-temporal expression pattern of cytoplasmic Cu/Zn superoxide dismutase (SOD1) mRNA during mouse embryogenesis. J Mol Histol. 2008;39:95–103. https://doi.org/10.1007/s10735-007-9134-1.
225. Zaken V, Kohen R, Ornoy A. The development of antioxidant defense mechanism in young rat embryos *in vivo* and *in vitro*. Early Pregnancy. 2000;4:110–23.
226. Zakeri ZF, Ahuja HS. Cell death/apoptosis: normal, chemically induced, and teratogenic effect. Mutat Res. 1997;396:149–61.

Chapter 3
Congenital Scoliosis and Segmentation Defects of the Vertebrae in the Genetic Clinic

Peter D. Turnpenny

The group of disorders that is best understood in terms of causation and pathophysiology is dominated by the family of conditions known as the 'spondylocostal dysostoses' (SCD), where segmentation anomalies occur throughout the vertebral column. Beyond this, there are a very large number of rare syndromes that include SDV and for which the genetic basis has been elucidated through various genetic strategies, though the pathophysiology giving rise to SDV is often not understood.

Six Notch signalling pathway genes are now linked to autosomal recessive (AR) SCD, types 1–6, and one to autosomal dominant (AD) SCD – type 5. SCD1 is caused by mutated delta-like 3 (*DLL3*) at chromosome 19q13.1; SCD2, and the severe spondylo*thoracic* dysostosis (STD), is due to mutated mesoderm posterior 2 (*MESP2*) at 15q26; SCD3 is due to mutated LFNG O-fucosylpeptide 3-beta-N-acetylglucosaminyltransferase (*LFNG*) at 7p22; and SCD4 is due to mutated hairy and enhancer of split 7 (*HES7*) gene at 17p13.2. SCD5, following autosomal dominant (AD) inheritance, is due to mutated *T-box 6* (*TBX6)* at 16p11.2, but bi-allelic mutations of *TBX6* also give rise to a variety of phenotypes ranging from congenital scoliosis to SCD. SCD6 is due to mutated transcriptional repressor *RIPPLY2* at 6q14.2, though in the first report the phenotype is distinct from other forms of SCD with segmentation defects affecting the posterior elements of C1–C4, with hemivertebrae and butterfly vertebrae of T2–T7.

Klippel-Feil syndrome (KFS), characterized by fusion of the cervical vertebrae, also embraces much diversity. KFS1 and KFS3 are AD forms and due to mutated *GDF6* (8p22.1) and *GDF3* (12p13.3), respectively. KFS2 is AR and due to mutated *MEOX1* (17q21). Vertebral segmentation anomalies are a variable feature of a wide variety of rare syndromes, but for a high proportion of the diverse radiological and developmentally aberrant phenotypes seen in clinical practice, the underlying cause

P. D. Turnpenny (✉)
Clinical Genetics Department, Royal Devon & Exeter NHS Foundation Trust, Exeter, UK

University of Exeter Medical School, Exeter, UK
e-mail: peter.turnpenny@nhs.net

K. Kusumi, S. L. Dunwoodie (eds.), *The Genetics and Development of Scoliosis*, https://doi.org/10.1007/978-3-319-90149-7_3

is unknown. Further progress will depend on identifying causative genes in familial cases of CS/SDV, or cohorts of subjects with similar phenotypes, using next-generation DNA sequencing. Several classifications for SDV, CS, and KFS have been proposed and they are described.

Introduction

Prior to the molecular genetic era, our understanding of genetic risk was often derived from epidemiological studies, leading to the assembly of empiric data that would allow analysis of possible inheritance patterns and recurrence risks for the purpose of genetic counselling. In 134 infants with idiopathic scoliosis, and their first-degree relatives, Wynne-Davies (1975) [1] found approximately 3% of parents and 3% of siblings had the same, or a similar, deformity. Congenital heart disease occurred in 2.5% of these cases (population incidence ~6/1000 live births) and intellectual disability in 13%, strongly suggesting an admixture of syndromic forms of CS. Genitourinary abnormalities were reported by Vitko et al. (1972) [2] in 37 of 85 (43%) patients with CS, and Erol et al. (2004) [3] studied 81 patients with different forms of CS and SDV, 39 of whom were prospectively recruited and 15 (38%) were found to have multi-organ/syndromic associations, many of which fitted loosely into the oculo-auriculo-vertebral (OAV) (or Goldenhar) spectrum. Purkiss et al. (2002) [4] studied 237 cases of congenital scoliosis and identified 49 where two or more family members had either congenital or idiopathic scoliosis, suggesting a much higher recurrence rate of 20.7%. There was also a history of idiopathic scoliosis in 17.3% of the families. Maisenbacher et al. (2005) [5] reported that 10% of congenital scoliosis cases described having first-degree relatives with idiopathic scoliosis. These risk data are diverse, and there is a need for more studies with clearer phenotypic stratification.

Table 3.1 lists rare syndromes that may include CS and/or SDV, along with the genetic basis, if known (possible associations also listed [6, 7]). Most are rare, and, where the genetic basis of these rare syndromes with segmentation anomalies is known, the mechanisms leading to abnormal vertebral formation are usually not elucidated. The most commonly encountered diagnostic groups in clinical practice are OAV/Goldenhar spectrum, VATER or VACTERL (vertebral, anal, cardiac, tracheo-esophageal, renal, and limb) association, MURCS (Müllerian duct aplasia, renal aplasia, cervicothoracic somite dysplasia) association, and maternal diabetes syndrome. The pathogenesis of these broad clinical groups is also poorly understood. OAV spectrum disorders, for which there is no clear consensus regarding diagnostic criteria, has been investigated using whole exome sequencing. Variants in *MYT1*, which encodes the myelin transcription factor, were found in occasional cases of OAVS, the mechanism possibly involving the retinoic acid receptor β [8]. A recent review has highlighted clinical heterogeneity in OAVS and collated reports

Table 3.1 Some syndromes and disorders that include segmentation defects of the vertebrae

Syndromes/disorders	OMIM	Gene
Alagille syndrome	118,450	*JAG1, NOTCH2*
Atelosteogenesis type II (de la Chapelle dysplasia)	256,050	*SLC26A2*
Atelosteogenesis type III	108,721	*FLNB*
Campomelic dysplasia	114,290	*SOX9*
Casamassima-Morton-Nance syndrome	271,520	
Caudal dysgenesis syndrome	600,145	*VANGL1*
Cerebro-facio-thoracic dysplasia	213,980	*TMCO1*
CHARGE syndrome	214,800	*CHD7*
'Chromosomal abnormality'		
Cleft-limb-heart malformation syndrome	215,850	
Currarino syndrome	176,450	*MNX1*
22q11.2 deletion syndrome (DiGeorge syndrome / velocardiofacial syndrome)	188,400; 192,430	
Dyssegmental dysplasia, Rolland-Desbuquois type	224,400	
Dyssegmental dysplasia, Silverman-Handmaker type	224,410	*HSPG2*
Facial dysmorphism with multiple malformations	227,255	
Femoral hypoplasia-unusual facies syndrome	134,780	
Fibrodysplasia ossificans progressiva	135,100	*ACVR1*
Goldenhar syndrome / Oculo-auriculo-vertebral (OAV) spectrum	164,210	*MYT1*
Incontinentia pigmenti	308,300	*IKBKG*
Kabuki syndrome	147,920	*KMT2D, KDM6A*
McKusick-Kaufman syndrome	236,700	*MKKS*
KBG syndrome	148,050	*ANKRD11*
Klippel-Feil syndrome	118,100	*GDF6, GDF3, MEOX1*
Larsen syndrome	150,250	*FLNB*
Lower mesodermal agenesis		
Maternal diabetes mellitus		
Mayer-Rokitansky-Kuster-Hauser syndrome	277,000	*TBX6, WNT4 , WNT9B*
MURCS association	601,076	*TBX6*
Multiple pterygium syndrome, Escobar variant	265,000	*CHRNG*
OEIS complex	258,040	
Phaver syndrome	261,575	
Postaxial acrofacial dysostosis	263,750	*DHODH*
RAPADILINO syndrome (*RECQL4*-related disorders)	266,280	*RECQL4*
Robinow syndrome – Autosomal dominant, *WNT5A*-related	180,700	*WNT5A*
Robinow syndrome – Autosomal recessive, *ROR2*-related	268,310	*ROR2*
Simpson-Golabi-Behmel syndrome type 1	312,870	*GPC3, GPC4*
Spinal dysplasia, Anhalt type	601,344	
Spondylocarpotarsal synostosis syndrome	272,460	*FLNB*

(continued)

Table 3.1 (continued)

Syndromes/disorders	OMIM	Gene
Urioste – Limb deficiency-vertebral hypersegmentation-absent thymus [1]		
VATER / VACTERL	192,350	
Verheij syndrome	615,583	*PUF60*
Wildervanck syndrome	314,600	

associating the phenotype with diverse chromosome and microarray-CGH findings, for which there is no consistent pattern [9]. Similarly, in 115 VATER/VACTERL subjects, a diverse range of copy number imbalances were found in 20 cases without a clear pattern of causality emerging [10], and there has been speculation that this group of disorders is a form of laterality defect [11]. There has been more significant progress with MURCS, as discussed below. Emerging from the confusing diversity of phenotypes, next-generation sequencing has also facilitated identification of at least one new syndrome, namely, that first reported by Verheij et al. [12], due to variants in *PUF60* [13], which includes cervical spine anomalies.

Any case series presenting to the spinal surgeon and/or paediatrician/geneticist will demonstrate enormous radiological and structural heterogeneity, and a syndromic or genetic diagnosis will often be imprecise at best and completely elusive at worst. Table 3.1 highlights that young (and not so young) patients presenting with CS/SDV should be examined and investigated very thoroughly for additional anomalies and a syndrome diagnosis considered. Referral to a clinical geneticist should therefore be part of the patient care pathway, and in the genetic clinic, investigations will almost certainly include microarray-CGH and, increasingly, analysis using a gene panel bespoke for skeletal dysplasias, abnormal segmentation disorders, and/or relevant dysmorphic/complex syndromes.

Although CS is frequently associated with SDV, this is not always so and CS may occur in the absence of segmentation anomalies, though abnormalities of vertebral *formation* may be present. In cases of this kind, a diagnosis of one of the skeletal dysplasias should be considered, though a precise radiological diagnosis may require follow-up skeletal surveys as the child grows. A clinical genetics opinion with a view to genetic testing may be very helpful, and examples include congenital contractural arachnodactyly (*aka* Beals syndrome), which is autosomal dominant and due to mutations in *FBN2*; chondrodysplasia punctata, Conradi-Hünermann type (*aka* Happle syndrome), which is X-linked and due to mutations in the *EBP* gene; diastrophic dysplasia, which is autosomal recessive and due to mutations in the sulphate transporter gene *SLC26A2* (*aka DTDST*); and spondylometaphyseal dysplasia, Kozlowski type, which is autosomal dominant and due to mutations in *TRPV4*.

Spondylocostal Dysostosis, Somitogenesis, and the Notch Signalling Pathway

The main progress in understanding the genetic basis of SDV has come through the study of somitogenesis in animal models, mainly mouse but also chick. Animals with specific gene knockouts are generated and multiple gene expression assays undertaken to help elucidate the developmental pathways. Somitogenesis is the sequential process whereby paired blocks of paraxial mesoderm are patterned and laid down on either side of the midline from the presomitic mesoderm (PSM) to form somites, a process that takes place between days 20 and 32 of human embryonic development, proceeding in a rostro-caudal direction. In mouse, a pair of somites is formed every 1–3 h, whilst in humans the process is estimated to take 6–12 h based on cell culture models and analysis of staged anatomical collections [14, 15]. Somites ultimately give rise to four substructures – sclerotome, which forms the axial skeleton and ribs; dermatome, which forms the dermis; myotome, which forms the axial musculature; and syndetome, which forms the tendons [16, 17]. Somitogenesis begins shortly after gastrulation and continues until the preprogrammed number of somite blocks is formed. In man 31 blocks of paired tissue are formed, but the number is species-specific. The establishment of somite boundaries takes place as a result of very finely tuned molecular processes determined by activation and negative feedback interactions between components of the Notch, Wnt, and FGF signalling pathways [18, 19] (Fig. 3.1). In the rostral third of the PSM, formation of segmental boundaries is subject to levels of the morphogen FGF8, which is produced in the caudal region of the embryo [20] and which probably maintains cells in an immature state until levels fall below a threshold, allowing boundary formation. Somites already harbour specification towards their eventual vertebral identity, a process regulated by the Hox family of transcription factors [21], which also display oscillatory expression in the mouse during somitogenesis [22].

The Wnt signalling pathway also displays oscillatory expression in a different temporal phase from Notch pathway genes and plays a key role in the segmentation clock [23–25]. The mediators of the determination front and the segmentation clock (Notch, FGF, Wnt) are required to form the somite boundary and specify rostrocaudal patterning of presumptive somites, for which *Mesp2* is crucial [26]. *Mesp2* is expressed caudal to the somite which is forming and this domain is set where Notch signalling is active, FGF signalling is absent, and the transcription factor *Tbx6* is expressed. Precise periodicity in the establishment of somite blocks is mediated by several so-called 'cycling', or 'oscillatory', genes, two of which, *LFNG* and *HES7*, are implicated in human SCD.

Somites themselves, having formed, are subsequently partitioned into rostral and caudal compartments, with vertebrae formed from the caudal compartment of

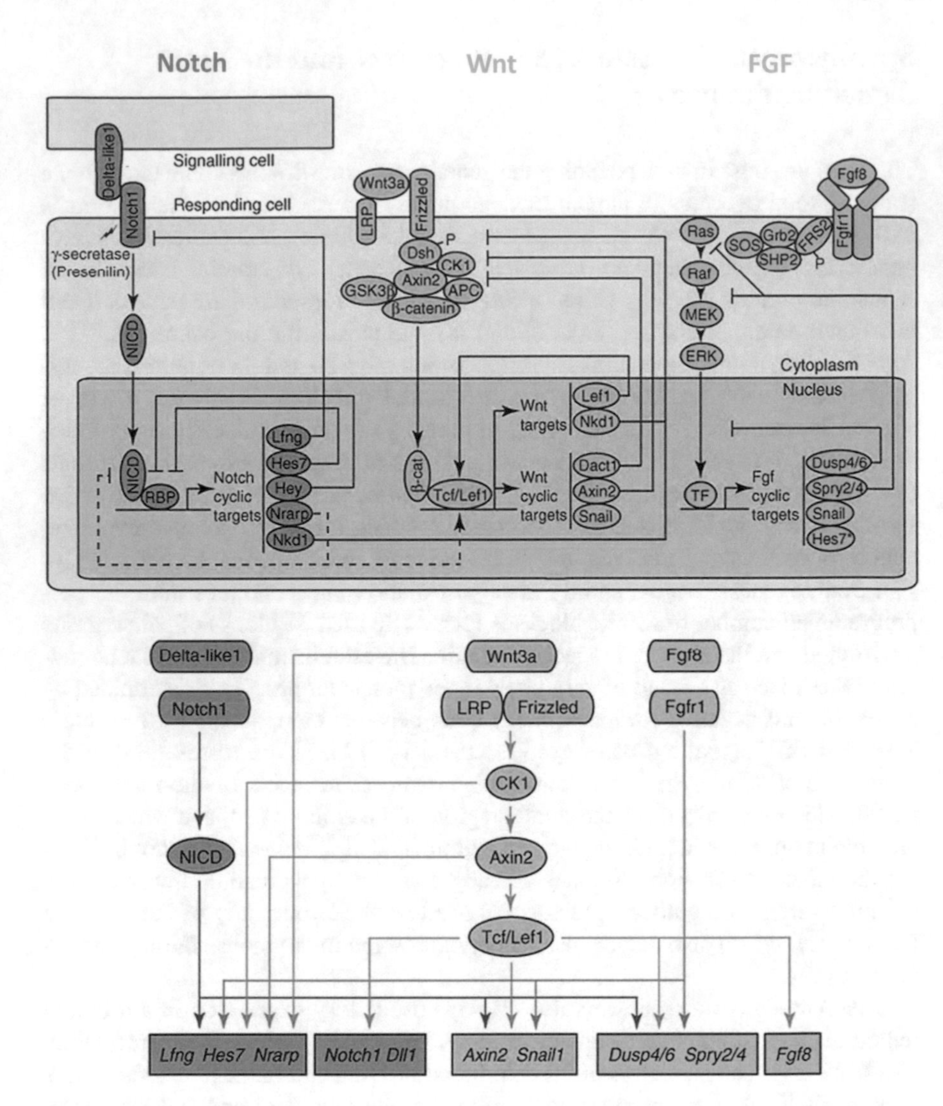

Fig. 3.1 The putative relationships between the Notch, Wnt, and FGF pathways in somitogenesis. (Reproduced courtesy of Elsevier; Gibb et al. [35])

one somite and the adjacent rostral compartment of the next, a phenomenon that is known as 'resegmentation' [27–30]. An understanding of the molecular biology of somitogenesis in animal models, in combination with finding patients and families with specific forms, or patterns, of segmentation anomalies, has led to the most definitive progress in understanding the causes of rare Mendelian forms of SDV.

Varied Use of Clinical Terminology

Spondylocostal Dysostosis

In clinical practice the use of terms for vertebral segmentation abnormalities has been inconsistent and confusing. 'Spondylocostal dysostosis' (SCD) continues to be applied to a wide variety of radiological phenotypes where abnormal segmentation is evident, together with rib involvement. For this review we use our preferred definition as given in Table 3.2. This restricts use of the term to *generalized* SDV, which defines the Mendelian forms of SCD thus far identified, as summarized in Table 3.3. This is usually a short trunk, short stature condition with multiple/generalized vertebral SDV accompanied by rib fusions and/or malalignment. A mild, non-progressive kyphoscoliosis is present, usually without additional organ abnormalities. Six Notch signalling pathway genes are now linked to this group, all demonstrating autosomal recessive (AR) inheritance with *TBX6* also demonstrating autosomal dominant (AD) inheritance, as described below.

Table 3.2 Proposed definitions for the terms spondylocostal dysostosis (SCD) and spondylothoracic dysostosis (STD) (ICVAS)

Features	Spondylocostal dysostosis (SCD)	Spondylothoracic dysostosis (STD)
General	No major asymmetry to chest shape Mild, non-progressive scoliosis Multiple SDV (M-SDV) ≥10 contiguous segments Absence of a bar Malaligned ribs with intercostal points of fusion	Chest shape symmetrical, with ribs fanning out in a 'crab-like' appearance Mild, non-progressive scoliosis or no scoliosis Generalized SDV (G-SDV) Regularly aligned ribs, fused posteriorly at the costovertebral origins, but no points of intercostal fusion
Specific, descriptive	'Pebble beach' appearance of vertebrae in early childhood radiographs (Fig. 3.3)	'Tramline' appearance of prominent vertebral pedicles in early childhood radiographs, not seen in SCD (Fig. 3.6) 'Sickle cell' appearance of vertebrae on transverse imaging (Cornier et al. [55])

Table 3.3 Genes causing generalized SDV, i.e. 'spondylocostal dysostosis' according to the definition proposed in Table 2

SCD	Gene symbol	Chromosomal locus	Protein name
SCD type 1	*DLL3*	19q13	Delta-like protein 3
SCD type 2 *and* STD	*MESP2*	15q26.1	Mesoderm posterior protein 2
SCD type 3	*LFNG*	7p22	Beta-1,3-N-acetylglucosaminyltransferase lunatic fringe
SCD type 4	*HES7*	17p13.2	Transcription factor HES-7
SCD type 5	*TBX6*	16p11.2	T-box6 protein
SCD type 6	*RIPPLY2*	6q14.2	Ripply transcriptional repressor 2

Table 3.4 Classification of SDV according to Mortier et al. (1996) [25]

Nomenclature	Definition
Jarcho-Levin syndrome	Autosomal recessive Symmetrical crab-like chest, lethal
Spondylothoracic dysostosis	Autosomal recessive Intrafamilial variability, severe/lethal Associated anomalies uncommon
Spondylocostal dysostosis	Autosomal dominant Benign
Heterogeneous group	Sporadic Associated anomalies common

Table 3.5 Classification/definition of SDV according to Takikawa et al. (2006) [26]

Nomenclature	Definition
Jarcho-Levin syndrome	Symmetrical crab-like chest
Spondylocostal dysostosis	≥2 vertebral anomalies associated with rib anomalies (fusion and/or absence)

Table 3.6 Classification (surgical/anatomical) of vertebral segmentation abnormalities causing congenital kyphosis/kyphoscoliosis, according to McMaster and Singh (1999) [27]

Type	Anatomical deformity	Anomalies
I	Anterior failure of vertebral body formation	Posterolateral quadrant vertebrae Single vertebra Two adjacent vertebrae Posterior hemivertebrae Single vertebra Two adjacent vertebrae Butterfly (sagittal cleft) vertebrae Anterior or anterolateral wedged vertebrae Single vertebra Two adjacent vertebrae
II	Anterior failure of vertebral body segmentation	Anterior unsegmented bar Anterolateral unsegmented bar
III	Mixed	Anterolateral unsegmented bar contralateral posterolateral quadrant vertebrae
IV	Unclassifiable	

A number of attempts have been made to classify SDV. The scheme proposed by Mortier et al. (1996) [31] combines phenotype and inheritance pattern (Table 3.4). The scheme proposed by Takikawa et al. (2006) [32] allows a very broad definition of SCD (Table 3.5), and both these schemes identify Jarcho-Levin syndrome (JLS) with a 'crab-like' chest. McMaster and Singh's [33] surgical approach to classification (1999) distinguishes between formation and segmentation errors (Table 3.6). As with McMaster's scheme, Aburakawa's [34] classification scheme for vertebral abnormalities (1996), which includes vertebral morphology (Table 3.7), does not attempt to identify phenotypic patterns of malformation based on assessment of the

Table 3.7 Aburakawa classification of vertebral segmentation abnormalities (Aburakawa et al. 1996 [28]; Takikawa et al. 2006 [26]) (modified North American classification). Note that hemivertebrae are seen in types B to F and L

Failure of formation
Type I
A. Double pedicle
B. Semi-segmented
C. Incarcerated
Type II
D. Non-incarcerated, no lateral shift
E. Non-incarcerated, plus lateral shift
Type III
F. Multiple
Type IV
G. Wedge
H. Butterfly
Failure of segmentation
I. Unilateral bar
J. Complete block
K. Wedge (plus narrow disc)
Mixed
L. Unilateral bar plus hemivertebrae
M. Unclassifiable

spine as a whole. The use of a limited number of terms in these classification schemes neither reflects the great diversity of radiological SDV phenotypes seen in clinical practice nor incorporates knowledge from molecular genetics. Furthermore, the diversity of SDV is not fully captured within the classification of osteochondrodysplasias [35, 36]. A scheme for classification and reporting from the International Consortium for Vertebral Anomalies and Scoliosis (ICVAS) is described later.

Klippel-Feil Syndrome

The term Klippel-Feil anomaly, or syndrome (KFS), has a more specific application, even though the phenotypes within the general category are diverse. KFS refers to vertebral fusion or segmentation errors involving the cervical region and has been the subject of several classifications (Table 3.8) [37, 38]. Clarke et al. (1998) [39] (Table 3.9) proposed a further, detailed classification combining modes of inheritance. To these clinical classifications must now be added a classification based on the recently discovered gene associations with rare forms of KFS [40–43] (Table 3.10). The *Pax1* gene has been shown to be active during sclerotome formation and differentiation, and mutations were identified in the mouse *undulated*, suggesting that sclerotome condensation is a *Pax1*-dependent process [44]. Two studies on patient cohorts with KFS were subsequently undertaken [6, 45], but despite some gene variants being identified in a small number, the same variants were either detected in an asymptomatic parent or did not occur in a conserved region of the gene. Overall, the role of *PAX1* in KFS remains to be elucidated.

Table 3.8 Classification of Klippel-Feil anomaly, referring to segmentation defects or fusion of the cervical vertebrae, according to Feil (1919) [31] and Thomsen et al. (1997) [32]

Type	Site	Anomaly
I	Cervical and upper thoracic	Massive fusion with synostosis
II	Cervical	One or two interspaces only, hemivertebrae, occipito-atlanto fusion
III	Cervical and lower thoracic or lumbar	Fusion

Table 3.9 Classification of Klippel-Feil anomaly according to Clarke et al. (1998) [33] (adapted from original publication)

Class	Vertebral fusions	Inheritance	Possible anomalies
KF1	Only class with C1 fusions C1 fusion not dominant Variable expression of other fusions	Recessive	Very short neck; heart; urogenital; craniofacial; hearing; limb; digital; ocular defects Variable expression
KF2	C2–3 fusion dominant C2–3 most rostral fusion Cervical, thoracic, and lumbar fusion variable within a family	Dominant	Craniofacial; hearing; otolaryngeal; skeletal and limb defects Variable expression
KF3	Isolated cervical fusions Variable position Any cervical fusion except C1	Recessive or reduced penetrance	Craniofacial Facial dysmorphology Variable expression
KF4	Fusion of cervical vertebrae, data limited	Possible X-linked Predominantly females	Hearing and ocular anomalies – abducens palsy with retraction bulbi *aka* Wildervanck syndrome

Table 3.10 Genes associated with Klippel-Feil syndrome (KFS) [34–36]

KFS	Gene symbol	Chromosomal locus	Encodes	Inheritance
KFS1	*GDF6* [*aka* cartilage-derived morphogenetic protein 2 (*CDMP2*)]	8q22.1	A member of the bone morphogenetic protein family	AD
KFS2	*MEOX1*	17q21.31	Homeodomain-containing protein	AR
KFS3	*GDF3*	12p13.1	A member of the bone morphogenetic protein family	AD
KFS4	*MYO18B*	22q12.1	Protein involved in muscle development	AR

The Genetics and Clinical Description of SCD Subtypes

SCD1

Autozygosity mapping was used to identify a locus for AR SCD at chromosome 19q13.1 in a large Arab–Israeli kindred first reported in 1991 [46, 47]. The region is syntenic with murine chromosome 7 harbouring the *Dll3* gene, which is truncated in the pudgy mouse [48]. *Dll3*-null mice show disruption of the cyclical somitogenesis clock within the PSM [49]. Human *DLL3* was therefore the obvious candidate for SCD1 in three separate families where affected individuals were homozygous for mutations [50]. The organization of human *DLL3* is almost identical to mouse *Dll3*, except the terminal exon, which corresponds to a fusion of mouse exons 9 and 10, resulting in a human protein of 32 additional amino acids. There is variability in the size of the mouse and human introns. The gene is sequentially ordered with a signal sequence (SS), the delta-serrate-lag (DSL) domain, six highly conserved epidermal growth factor (EGF) repeats, and a transmembrane (TM) region (Fig. 3.2). More than 30 *DLL3* mutations have now been identified and most of these published [51, 52]. Approximately 75% of positive cases have protein truncation nonsense mutations (the rest being missense), and parental consanguinity is seen in roughly the same proportion of cases.

There is a general consistency in the abnormal form and shape of the vertebrae in the different regions, from cervical to lumbar. The radiological appearances in childhood are of vertebrae that are circular or ovoid on A-P projection, and they have smooth outlines. To this appearance the term 'pebble beach sign' has been applied [52] (Fig. 3.3). Stature is affected to a variable degree, with some affected subjects achieving a final adult height that is only about 15 cm less than their predicted height on the basis of arm span measurements (assuming arm length is unaffected). Final adult stature is more severely affected in some cases, and in the large family reported by Turnpenny et al. (1991) [46], a range of severity was evident. We know of two patients with slightly milder phenotypes due to missense mutations C309R and G404C, both with less dramatic vertebral segmentation abnormalities, even though the whole spine was involved (unpublished data). The milder phenotype may be due to the position of these residues within the EGF domains. It seems

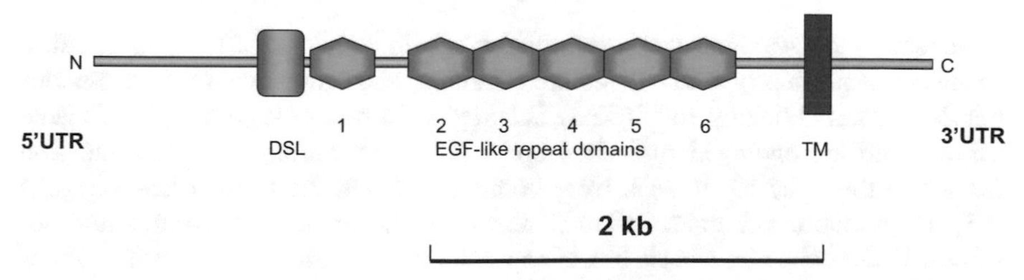

Fig. 3.2 The organization of the *DLL3* gene. DSL delta-serrate-lag domain, EGF epidermal growth factor domain(s), SS starter sequence, TM transmembrane domain

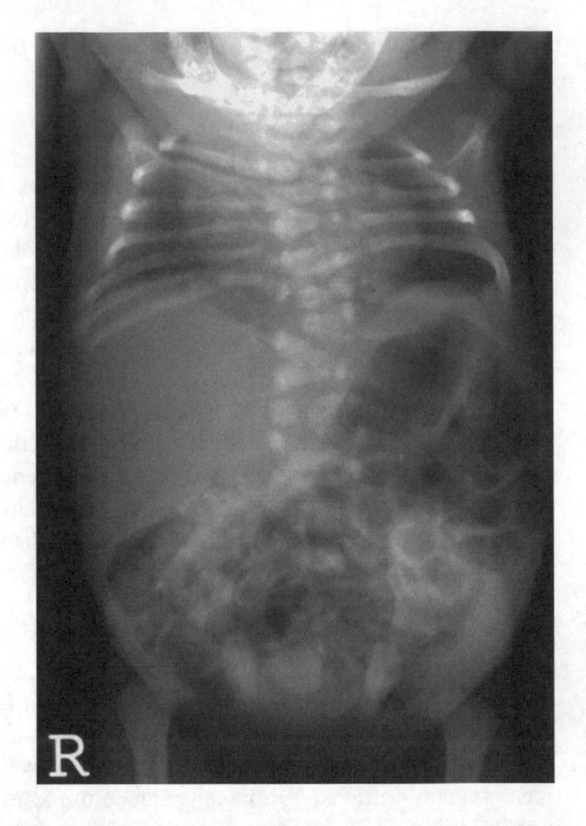

Fig. 3.3 The radiological phenotype of SCD1 due to mutated *DLL3*. This shows segmentation abnormalities throughout the vertebral column and the variable ovoid appearance of multiple vertebrae – the 'pebble beach' sign. The ribs are malaligned with points of fusion along their length. (Reproduced courtesy of GeneReviews)

likely that some missense mutations, though not all, may give rise to milder phenotypes. There is no clear, consistent evidence for organ abnormalities beyond the spine in subjects with SCD1. Learning difficulties or intellectual disability is not a feature, and although affected individuals have mild scoliotic curves from an early stage, these appear to remain stable throughout life in the majority of cases, and spinal surgery is usually not required.

SCD2

A genome-wide scanning was used to identify linkage to 15q21.3–15q26.1 in a consanguineous family with two affected children who neither had *DLL3* mutations nor demonstrated linkage to 19q13.1. This region harbours the somitogenesis gene *MESP2*, and sequencing identified a 4-bp (AGGC) duplication, frameshift mutation for which the affected subjects were homozygous and the parents heterozygous [53]. The mutation was not found in 68 normal ethnically matched control chromosomes. *MESP2* encodes a basic helix-loop-helix (bHLH) transcription factor, a protein of 397 amino acids. Human MESP2 protein shares 58.1% identity with mouse Mesp2. Human *MESP2* amino terminus contains a bHLH region encompassing 51

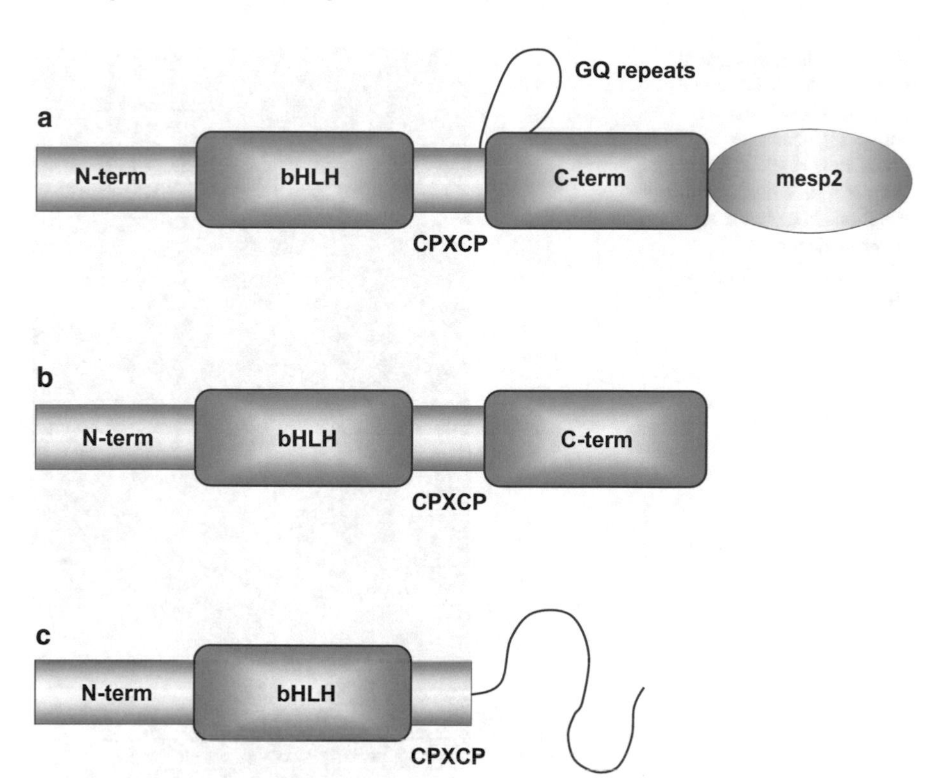

Fig. 3.4 Comparison of mesoderm posterior 2: (**a**) *MESP2* and (**b**) *MESP1* (adjacent to *MESP2* at 15q26 in human). Both sequences clearly contain a basic helix–loop–helix (bHLH) domain. The length of the loop region is conserved between *MESP1* and *MESP2*. In addition, *MESP1* and *MESP2* contain a unique CPXCP motif immediately C-terminal to the bHLH domain. *MESP1* and *MESP2* also share a C-terminal region that is likely to adopt a similar fold. *MESP2* sequences contain a unique region at the C-terminus. The GQ repeats, which are located between the CPXCP motif and the shared C-terminal domain, are also found in human *MESP1* (only two repeats) but have expanded in human *MESP2* (13 repeats). Although lacking GQ repeats, the mouse sequences have two QX repeats in the same region: mouse *MESP1* QSQS and mouse *MESP2* QAQM. (**c**) Human *MESP2* frameshift mutant

amino acids, which is divided into an 11-residue basic domain, a 13-residue helix I domain, an 11-residue loop domain, and a 16-residue helix II domain. The loop region is conserved between mouse and human *Mesp1* and *Mesp2*. In addition, both *Mesp1* and *Mesp2* contain a unique CPXCP motif immediately carboxy-terminal to the bHLH domain. The amino- and carboxy-terminal domains are separated in human *MESP2* by a GQ repeat region, which is also present in human *MESP1* (2 repeats), but have expanded in human *MESP2* (13 repeats). Mice *Mesp1* and *Mesp2* do not contain any GQ repeats, but they do contain a couple of QX repeats in the same region (Fig. 3.4). In cases designated SCD2, the mutations identified do not appear to give rise to nonsense-mediated decay of the derivative protein, in contrast to the effect of mutations in STD (see below), the more severe phenotype due to mutated *MESP2*.

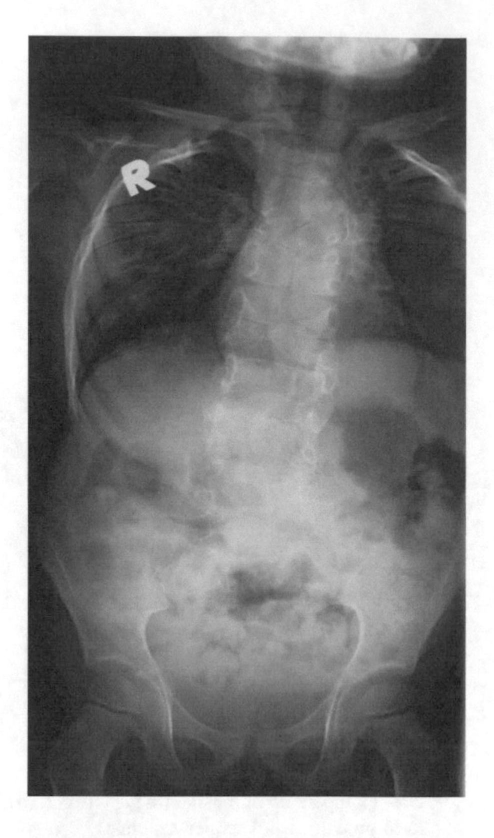

Fig. 3.5 The radiological phenotype of SCD2 due to mutated *MESP2*. This shows segmentation abnormalities throughout the vertebral column with the thoracic region most severely disrupted. (Reproduced courtesy of GeneReviews)

Only one family with SCD due to a mutation in *MESP2*, demonstrating AR inheritance, has been published [53], and, therefore, the phenotype is based on minimal data. However, a second affected family with the same mutation and a very similar radiological phenotype was presented at an international meeting in 2005 (Bonafé et al.). Subsequent haplotype analysis failed to show evidence for a common ancestry for the two families (unpublished data), so the particular 4-bp duplication mutation is recurrent. A further case was found to be a compound heterozygote for *MESP2* mutations (Fig. 3.5). The radiological phenotype is similar to, but distinguishable from, that of SCD1, and the ribs are more normally aligned. Segmentation defects appear more severe in the thoracic vertebrae compared to the lumbar vertebrae, which are relatively spared. Stature is affected to a small degree, and no additional organ abnormalities have been reported.

STD: Spondylothoracic Dysostosis

Mutated *MESP2* is also the cause of STD [54], a severe form of SDV with marked shortening of the spine, reduced thoracic volume, and in some cases life-threatening respiratory insufficiency. The mutations are of the type that give rise to nonsense-medicated decay (unpublished data).

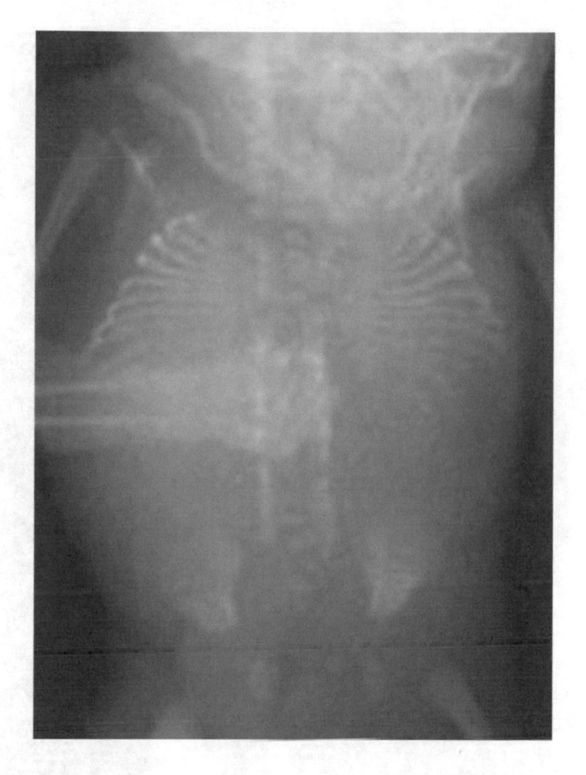

Fig. 3.6 The radiological phenotype of STD due to mutated *MESP2*. This shows marked shortening of the spine, generalized segmentation defects, a 'crab-like' fanning out of the ribs from their posterior costovertebral origins, and well-aligned ribs. The vertebral pedicles are ossified at this early stage of life (in contrast to SCD1), sometimes called the 'tramline' sign. (Reproduced courtesy of Wiley)

There are two very useful radiological features that help distinguish SCD type 1 from STD (and to some extent SCD2). Firstly, in SCD points of fusion of the ribs along their length are usually apparent, whereas in STD the ribs are fused *posteriorly* and fan out laterally ('crab-like' appearance) without points of fusion along their length. Secondly, in foetal life and early childhood, multiple rounded hemivertebrae characterize SCD1 (*DLL3* gene) – the 'pebble beach sign' [52] – and the vertebral pedicles are poorly visualized because they are not yet ossified. By contrast, in *MESP2*-associated SCD and STD, the vertebral pedicles are visible radiologically in foetal life and early childhood. Indeed, they are often neatly arranged and aligned, bordering multiple hemivertebrae between (Fig. 3.6). This is sometimes referred to as the 'tramline sign'. STD has been well delineated and described by Cornier et al. [55]. Most reported cases are Puerto Rican but, although rare, it has been seen globally.

SCD3

A candidate gene approach was used to identify *LFNG* as the genetic cause of SCD in an individual in whom no mutation could be found in *DLL3* or *MESP2* [56] (Fig. 3.7). *LFNG* – Lunatic fringe – encodes a glycosyltransferase (fucose-specific β-1, 3-N-acetylglucosamine) that post-translationally modifies the Notch family of

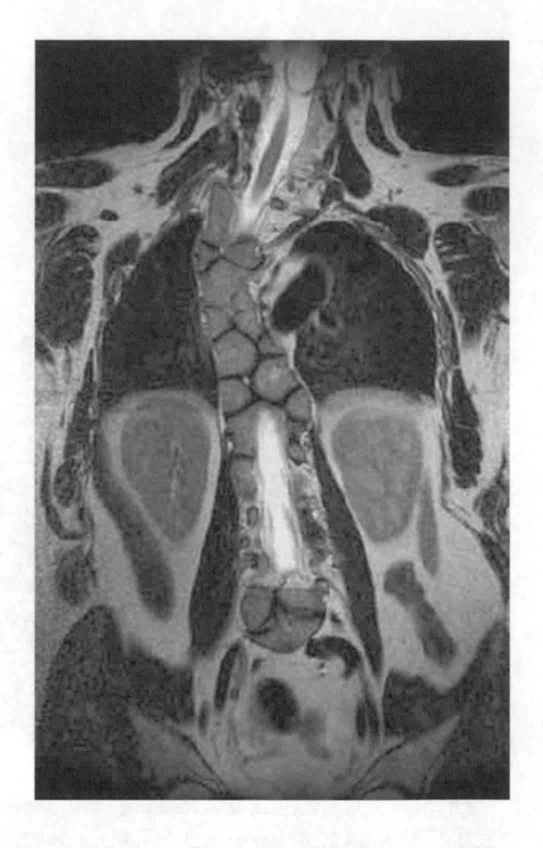

Fig. 3.7 The radiological phenotype of SCD3 due to mutated *LFNG*. This shows severe shortening of the thoracic spine in particular. (Reproduced courtesy of Elsevier)

cell surface receptors, a key step in the regulation of this signalling pathway [57], and is one of the 'cycling' genes whose wave of expression in the PSM, in a caudal-rostral direction, is crucial to the establishment of the next somite boundary. *LFNG* was sequenced as its expression is severely dysregulated in mouse embryos that lack *Dll3* [58, 59] (the phenotypes of *Dll3* and *Lfng* null mutant mice are very similar) and is associated with the Notch signalling pathway (like *DLL3* and *MESP2*). In the affected case a missense mutation (c.564C → A) was detected that resulted in the substitution of leucine for phenylalanine (F188 L). The proband's consanguineous parents, of Lebanese Arab origin, were normal and heterozygous for the mutant allele. Functional assays showed that F188 L did not localize to the Golgi apparatus as the wild-type LFNG protein and that F188 L lacked transferase activity.

In this case of SCD3, the segmentation disruption was severe compared to SCD1 and SCD2, giving rise to marked truncal shortening and apparently normal limb length – arm span 186.5 cm, adult height 155 cm, and lower segment 92.5 cm. Multiple vertebral ossification centres in the thoracic spine, with very angular shapes were apparent. The affected case also demonstrated a minor form of distal arthrogryposis in the upper limbs, and it is not known whether this was part of the condition or secondary to peripheral nerve entrapment.

SCD4

In humans, the first identified case of SCD due to a mutation in *HES7* was found to be homozygous for a C > T nucleotide transition in exon 2, resulting in an arginine to tryptophan amino acid substitution [60]. Subsequently a family demonstrating compound heterozygosity for *HES7* mutations was identified [61]. *HES7* encodes a bHLH-Orange domain transcriptional repressor protein that is both a direct target of the Notch signalling pathway and part of a negative feedback mechanism required to attenuate Notch signalling [62]. Like *LFNG*, *HES7* is a cycling gene; it is expressed in the PSM in an oscillatory pattern [63], which is achieved by an autoregulatory loop. Once translated, HES proteins act on their own promoters to repress transcription, and, due to the short half-life of HES proteins, autorepression is relieved, which allows a new wave of transcription and translation every 90–120 minutes in the mouse. Hes7-null mice display severe multiple SDV phenotypes [64].

To date, the pattern of SDV in SCD4 lies somewhere between SCD1 and mild STD, with ribs appearing to show fusion posteriorly and fanning out in a crab-like fashion (Fig. 3.8). The first reported patient was homozygous for a *HES7* mutation and also had a lumbar myelomeningocele neural tube defect [60], whilst there were no associated malformations in the second reported family [61], where SCD occurred only in subjects who were compound heterozygotes for *HES7* mutations. The findings in a large extended kindred, in which some family affected individuals have midline developmental defects besides generalized SDV, support the possibility that *HES7* is implicated in determination of laterality and neural tube closure [65].

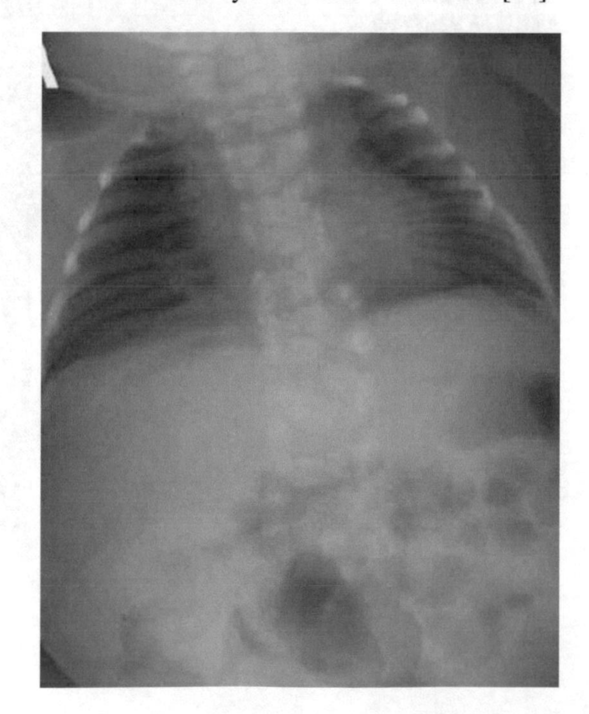

Fig. 3.8 The radiological phenotype of SCD4 due to mutated *HES7*. This shows segmentation abnormalities throughout the vertebral column, and the appearance resembles that of SCD2/STD. (Reproduced courtesy of Oxford University Press)

SCD5: Autosomal Dominant

Only one genetic cause of AD SCD has thus far been identified, namely, mutated *TBX6* [66], in a three-generation Macedonian family previously shown not having a mutation in *DLL3*, *MESP2*, *LFNG*, and *HES7* [67]. Exome capture and next-generation sequencing were used to identify a stop-loss mutation in *TBX6* that segregated with the phenotype in two generations, and the family demonstrated a generalized pattern of SDV without any additional malformations (Fig. 3.9).

The *TBX6*, or T-box6, gene encodes a putative DNA-binding protein expressed in somite precursor cells, indicating that it is implicated in the specification of the paraxial mesoderm. Studies in mouse demonstrate that the Tbx6 protein is directly bound to the *Mesp2* gene, mediates Notch signalling, and subsequent *Mesp2* transcription in the PSM [68]. Functional studies in this reported family [66] demonstrated a deleterious effect on the transcriptional activation activity of the TBX6 protein, probably secondary to haploinsufficiency.

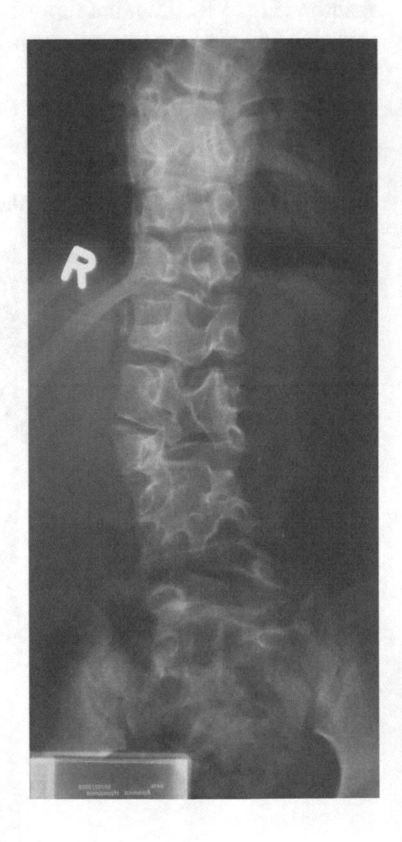

Fig. 3.9 The radiological phenotype of SCD5 due to mutated *TBX6*. This shows segmentation abnormalities throughout the vertebral column in an adult. The pattern is similar to that seen in an adult with SCD1. (Reproduced courtesy of Wiley)

SCD6

SCD6 is due to mutated transcriptional repressor *RIPPLY2* at 6q14.2 [69]. This is another rare condition with only one reported family thus far. Two brothers born to non-consanguineous parents had SDV involving the posterior elements of the cervical vertebrae, C1-C4, as well as hemivertebrae and butterfly vertebrae of T2-T7. Spinal cord compression appeared to be a consequence of kyphosis at C2-C3, and mild thoracic scoliosis was present [69]. The radiological pattern was distinct from other forms of SCD, and it can be argued that the condition is best classified elsewhere, especially in view of the overlap with KFS.

TBX6, Congenital Scoliosis, and MURCS

The most significant recent advance in this field is the elucidation of the role of various molecular events involving *TBX6*, and its locus at 16p11.2, in the development of both CS and SCD. In a large cohort of Han Chinese subjects with sporadic CS, Wu et al. (2015) [70] identified heterozygous null alleles in *TBX6* in 11%. Of this group, ~75% had a microdeletion at 16p11.2, whilst the remainder had nonsense or frameshift mutations. Deletion 16p11.2 is a well-recognized imbalance that confers susceptibility to neurodevelopmental problems, an increased head size, and mild obesity – but not all have CS. Wu et al. (2015) [70] therefore studied the other *TBX6* allele and identified a common haplotype in all affected cases. The haplotype is a series of single nucleotide variants, two of which, rs3809624 and rs3809627, are located in the 5′ noncoding region, and rs2289292 is a synonymous change in the final exon. This is known as the T-C-A haplotype as the transitions are C > **T**, T > **C**, and C > **A** reading from the 3′ end. The most common radiological phenotype is the presence of a limited number of hemivertebrae, most often in the upper lumbar or lower thoracic regions, giving rise to a varying degree of scoliosis. These findings have been confirmed elsewhere, and it is also now clear that a variety of axial skeletal phenotypes can result from these *TBX6* events, ranging from CS to SCD [71]. Our own unpublished work adds to these findings, including the finding of abnormal *TBX6* alleles in causing a form of MURCS, sometimes with a very severe axial skeletal pattern [72]. The molecular findings and phenotypes are summarized in Fig. 3.10.

A New Classification and Radiological Reporting System for SDV

Currently, the use of nomenclature to describe CS/SDV is inconsistent and confusing, even though some authors have recognized the existence of different entities and applied a rational distinction in the use of terms [73, 74]. This applies to the eponym *Jarcho-Levin* syndrome (JLS), used so widely that it has lost any

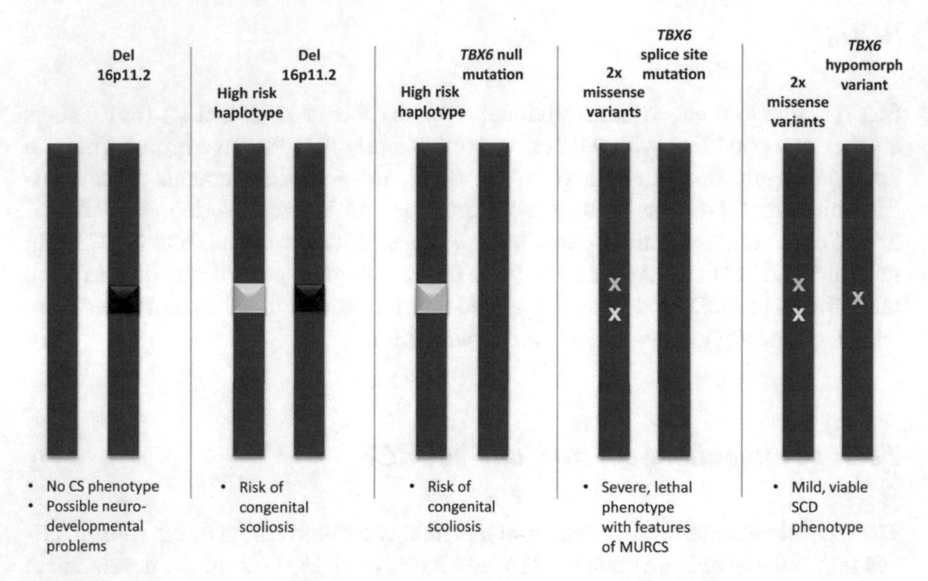

Fig. 3.10 A summary of molecular events and CS/SCD phenotypes involving *TBX6* and the 16p11.2 locus

specificity. Its use is therefore discouraged. *Klippel-Feil* anomaly, or syndrome, is long established, more specific, and therefore retains some usefulness. It is suggested that the terms SCD and STD be reserved for specific phenotypes (Tables 3.2 and 3.3). Strictly speaking, these are dysostoses, not dysplasias, because they are due to errors of segmentation or formation early in morphogenesis, rather than an ongoing abnormality of chondro-osseous tissues during pre- and postnatal life.

The widely used terms Jarcho-Levin syndrome (JLS) [75–80], costovertebral syndrome [81–83], spondylocostal dysostosis (SCD, or SCDO according to OMIM nomenclature) [84, 85], and spondylothoracic dysplasia (STD) [86–88], are used interchangeably and indiscriminately [32, 89, 90]. In 1938 Jarcho and Levin [91] reported two siblings with short trunks, multiple SDV (M-SDV), and abnormally aligned ribs with points of fusion. In recent years many authors have equated JLS with the distinctive phenotype of a severely shortened spine and a 'crab-like' appearance of the ribs for which the preferred term today is STD [54, 55], first suggested by Moseley & Bonforte (1969) [86]. Berdon et al. (2010) [92] have clarified the historical record. The incidence of STD is relatively high in Puerto Ricans compared to elsewhere because of a founder effect *MESP2* mutation [55]. The ethnicity of the siblings reported by Jarcho and Levin (1938) [91] was 'coloured', they did not manifest the distinctive crab-like appearance, and their phenotype was closer to either SCD2 or SCD4. A further eponym lacking specificity is Casamassima-Morton-Nance (CMN) syndrome [93]. This combines SDV with urogenital anomalies, apparently following autosomal recessive inheritance. However, subsequent reports [94, 95] demonstrated a different SDV phenotype from the cases of Casamassima et al. (1981) [82], and consistency across all three reports, based on the SDV phenotype, is poor.

The classification and reporting system for SDV conditions is illustrated in Fig. 3.11 and was developed by ICVAS. It provides simple, uniform terminology and can be applied both to man and animal models. The system takes account of both syndromic or non-syndromic conditions (see Table 3.1). Non-syndromic conditions include most cases of Mendelian SCD and STD (as defined in this paper), whereby the malformation is usually restricted to the spine. SDV may be single, multiple, show various regional involvement, be associated with kyphoscoliosis, and include KFS. These conditions are mainly caused by defective somitogenesis and/or non-intrinsic disruption of normal segmentation and/or formation of vertebrae. In the proposed scheme (Fig. 3.11), conditions essentially fall into one of seven categories. This simplification allows for uniformity between observers. In any particular case, once placed within one of the seven categories, further detailed descriptions of the position and effects of the segmentation anomalies can be added. Where appropriate, therefore, the ICVAS scheme incorporates existing terminology. This greatly reduces confusion that might be generated by indiscriminate use of the terms JLS or SCD. In cases with at least ten vertebral segments affected, but non-contiguously, we suggest the designation 'multi-regional' form of M-SDV, rather than 'generalized'. This group of phenotypes appears to be diverse, and further delineation will only be possible with advances in our understanding of causation.

The usefulness of correlating a detailed clinical examination with radiological findings has been well described previously [3]. The system was piloted [96] and allows for a more precise characterization of the radiological phenotype compared

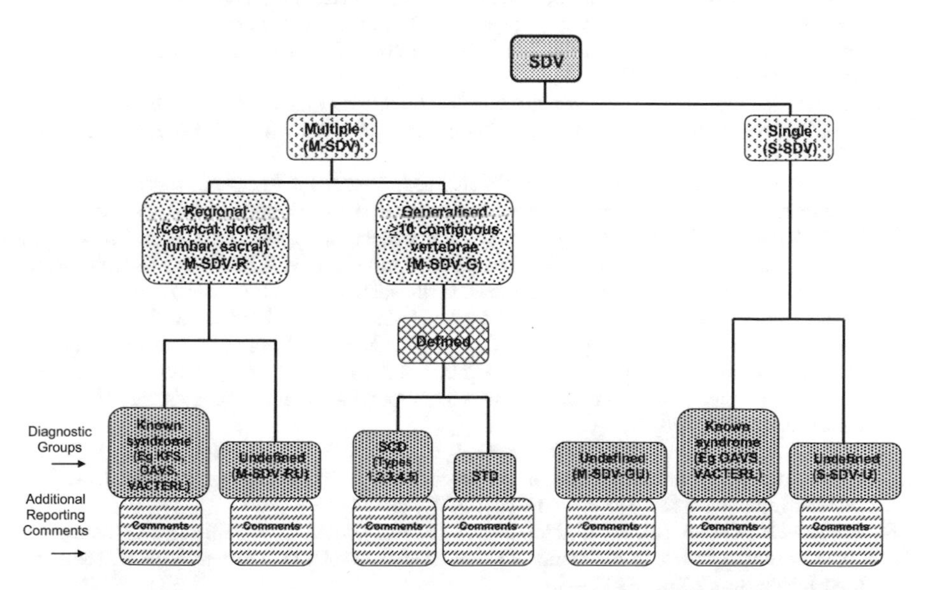

Fig. 3.11 The classification algorithm for SDV proposed by ICVAS, the International Consortium for Vertebral Anomalies and Scoliosis. By this scheme any patient can be placed within one of seven basic categories, with provision for further description of the specific findings in any individual. (ICVAS clinical classification algorithm. **SDV** Segmentation Defect(s) of the Vertebrae, **M** Multiple, **S** Single, **R** Regional, **G** Generalised, **U** Undefined)

to the indiscriminate use of a small number of terms, including eponyms. Furthermore, the system incorporates assessment of radiographic patterns of the spine as a whole, in addition to malformations of individual vertebrae. It is recognized that the system will evolve over time as the identification of new genes brings clarity to the causation of different conditions, and groups of conditions, with CS and SDV.

References

1. Wynne-Davies R. Infantile idiopathic scoliosis. Causative factors, particularly in the first six months of life. J Bone Joint Surg Br. 1975;57:138–41.
2. Vitko RJ, Cass AS, Winter RB. Anomalies of the genitourinary tract associated with congenital scoliosis and congenital kyphosis. J Urol. 1972;108:655–9.
3. Erol B, Tracy MR, Dormans JP, Zackai EH, Tonnesen M, O'Brien ML, et al. Congenital scoliosis and vertebral malformations: characterization of segmental defects for genetic analysis. J Pediatr Orthop. 2004;24:674–82.
4. Purkiss SB, Driscoll B, Cole WG, Alman B. Idiopathic scoliosis in families of children with congenital scoliosis. Clin Orthop Relat Res. 2002;401:27–31.
5. Maisenbacher MK, Han JS, O'Brien ML, Tracy MR, Erol B, Schaffer AA, et al. Molecular analysis of congenital scoliosis: a candidate gene approach. Hum Genet. 2005;116(5):416–9.
6. McGaughran J, Oates A, Donnai D, Read AP, Tassabehji M. Mutations in PAX1 may be associated with Klippel-Feil syndrome. Eur J Hum Genet. 2003;11:468–74.
7. Philibert P, Biason-Lauber A, Rouzier R, Pienkowski C, Paris F, Konrad D, et al. Identification and functional analysis of a new WNT4 gene mutation among 28 adolescent girls with primary amenorrhea and müllerian duct abnormalities: a French collaborative study. J Clin Endocrinol Metab. 2008;93(3):895–900.
8. Lopez E, Berenguer M, Tingaud-Sequeira A, Marlin S, Toutain A, Denoyelle F, et al. Mutations in MYT1, encoding the myelin transcription factor 1, are a rare cause of OAVS. J Med Genet. 2016;53:752–60.
9. Beleza-Meireles A, Clayton-Smith J, Saraiva JM, Tassabehji M. Oculo-auriculo-vertebral spectrum: a review of the literature and genetic update. J Med Genet. 2014;51:635–45.
10. Zhang R, Marsch F, Kause F, Degenhardt F, Schmiedeke E, Märzheuser S, et al. Array-based molecular karyotyping in 115 VATER/VACTERL and VATER/VACTERL-like patients identifies disease-causing copy number variations. Birth Defects Res. 2017;109:1063–9.
11. Sadler TW. Is VACTERL a laterality defect? Am J Med Genet A. 2015;167A:2563–5.
12. Verheij JB, de Munnik SA, Dijkhuizen T, de Leeuw N, Olde Weghuis D, van den Hoek GJ, et al. An 8.35 Mb overlapping interstitial deletion of 8q24 in two patients with coloboma, congenital heart defect, limb abnormalities, psychomotor retardation and convulsions. Eur J Med Genet. 2009;52(5):353–7.
13. Low KJ, Ansari M, Abou Jamra R, Clarke A, El Chehadeh S, FitzPatrick DR, et al. PUF60 variants cause a syndrome of ID, short stature, microcephaly, coloboma, craniofacial, cardiac, renal and spinal features. Eur J Hum Genet. 2017;25:552–9.
14. William DA, Gibson SB, JD TJ, Markov V, Gonzalez DM, et al. Identification of oscillatory genes in somitogenesis from functional genomic analysis of a human mesenchymal stem cell model. Dev Biol. 2007;305:172–86.
15. Eckalbar WL, Fisher RE, Rawls A, Kusumi K. Scoliosis and segmentation defects of the vertebrae. Wiley Interdiscip Rev Dev Biol. 2012;1(3):401–23.
16. Keynes RJ, Stern CD. Mechanisms of vertebrate segmentation. Development. 1988;103:413–29.
17. Brent AE, Schweitzer R, Tabin CJ. A somitic compartment of tendon progenitors. Cell. 2003;113(2):235–48.

18. Dequéant M-L, Pourquié O. Segmental patterning of the vertebrate embryonic axis. Nat Rev Genet. 2008;9:370–82.
19. Gibb S, Maroto M, Dale JK. The segmentation clock mechanism moves up a notch. Trends Cell Biol. 2010;20:593–600.
20. Dubrulle J, McGrew MJ, Pourquié O. FGF signaling controls somite boundary position and regulates segmentation clock control of spatiotemporal Hox gene activation. Cell. 2001;106:219–32.
21. Krumlauf R. Hox genes in vertebrate development. Cell. 1994;78:191–201.
22. Zákány J, Kmita M, Alarcon P, de la Pompa JL, Duboule D. Localized and transient transcription of Hox genes suggests a link between patterning and the segmentation clock. Cell. 2001;106(2):207–17.
23. Aulehla A, Wehrle C, Brand-Saberi B, Kemler R, Gossler A, Kanzler B, et al. Wnt3a plays a major role in the segmentation clock controlling somitogenesis. Dev Cell. 2003;4:395–406.
24. Aulehla A, Herrmann B. Segmentation in vertebrates: clock and gradient finally joined. Genes Dev. 2004;18:2060–7.
25. Hofmann M, Schuster-Gossler K, Watabe-Rudolph M, Aulehla A, Herrmann BG, Gossler A. WNT signaling, in synergy with T/TBX6, controls notch signaling by regulating Dll1 expression in the presomitic mesoderm of mouse embryos. Genes Dev. 2004;18(22):2712–7.
26. Saga Y. The mechanism of somite formation in mice. Curr Opin Genet Dev. 2012;22:331–8.
27. Remak R. Untersuchungen über die entwicklung der Wirbeltiere. Berlin: Reimer; 1850.
28. Bagnall KM, Higgins SJ, Sanders EJ. The contribution made by cells from a single somite to tissues within a body segment and assessment of their integration with similar cells from adjacent segments. Development. 1989;107(4):931–43.
29. Ewan KB, Everett AW. Evidence for resegmentation in the formation of the vertebral column using the novel approach of retroviral-mediated gene transfer. Exp Cell Res. 1992;198(2):315–20.
30. Goldstein RS, Kalcheim C. Determination of epithelial half somites in skeletal morphogenesis. Development. 1992;116:441–5.
31. Mortier GR, Lachman RS, Bocian M, Rimoin DL. Multiple vertebral segmentation defects: analysis of 26 new patients and review of the literature. Am J Med Genet. 1996;61:310–9.
32. Takikawa K, Haga N, Maruyama T, Nakatomi A, Kondoh T, Makita Y, et al. Spine and rib abnormalities and stature in spondylocostal dysostosis. Spine. 2006;31:E192–7.
33. McMaster MJ, Singh H. Natural history of congenital kyphosis and congenital kyphoscoliosis. A study of one hundred and twelve patients. J Bone Joint Surg Am. 1999;81:1367–83.
34. Aburakawa K, Harada M, Otake S. Clinical evaluations of the treatment of congenital scoliosis. Orthop Surg Trauma. 1996;39:55–62.
35. Offiah AC, Hall CM. Radiological diagnosis of the constitutional disorders of bone. As easy as a, B, C? Pediatr Radiol. 2003;33:153–61.
36. Bonafe L, Cormier-Daire V, Hall C, Lachman R, Mortier G, Mundlos S, et al. Nosology and classification of genetic skeletal disorders: 2015 revision. Am J Med Genet A. 2015;167A:2869–92.
37. Feil A. L'absence et la diminution des vertebres cervicales. Paris: Thesis, Libraire Litteraire et Medicale; 1919.
38. Thomsen M, Schneider U, Weber M, Johannisson R, Niethard F. Scoliosis and congenital anomalies associated with Klippel-Feil syndrome types I-III. Spine. 1997;22:396–401.
39. Clarke RA, Catalan G, Diwan AD, Kearsley JH. Heterogeneity in Klippel-Feil syndrome: a new classification. Pediatr Radiol. 1998;28:967–74.
40. Tassabehji M, Fang ZM, Hilton EN, McGaughran J, Zhao Z, de Bock CE, et al. Mutations in GDF6 are associated with vertebral segmentation defects in Klippel-Feil syndrome. Hum Mutat. 2008;29:1017–27.
41. Mohamed JY, Faqeih E, Alsiddiky A, Alshammari MJ, Ibrahim NA, Alkuraya FS. Mutations in MEOX1, encoding mesenchyme homeobox 1, cause Klippel-Feil anomaly. Am J Hum Genet. 2013;92:157–61.

42. Ye M, Berry-Wynne KM, Asai-Coakwell M, Sundaresan P, Footz T, French CR, et al. Mutation of the bone morphogenetic protein GDF3 causes ocular and skeletal anomalies. Hum Molec Genet. 2010;19:287–98.
43. Alazami AM, Kentab AY, Faqeih E, Mohamed JY, Alkhalidi H, Hijazi H, et al. A novel syndrome of Klippel-Feil anomaly, myopathy, and characteristic facies is linked to a null mutation in MYO18B. J Med Genet. 2015;52:400–4.
44. Balling R, Deutsch U, Gruss P. Undulated, a mutation affecting the development of the mouse skeleton, has a point mutation in the paired box of Pax 1. Cell. 1988;55:531–5.
45. Giampietro PF, Raggio CL, Reynolds CE, Shukla SK, McPherson E, Ghebranious N, et al. An analysis of PAX1 in the development of vertebral malformations. Clin Genet. 2005;68:448–53.
46. Turnpenny PD, Thwaites RJ, Boulos FN. Evidence for variable gene expression in a large inbred kindred with autosomal recessive spondylocostal dysostosis. J Med Genet. 1991;28:27–33.
47. Turnpenny PD, Bulman MP, Frayling TM, Abu-Nasra TK, Garrett C, Hattersley AT, et al. A gene for autosomal recessive spondylocostal dysostosis maps to 19q13.1-q13.3. Am J Hum Genet. 1999;65:175–82.
48. Kusumi K, Sun ES, Kerrebrock AW, Bronson RT, Chi DC, Bulotsky MS, et al. The mouse pudgy mutation disrupts Delta homolog Dll3 and initiation of early somite boundaries. Nature Genet. 1998;19:274–8.
49. Dunwoodie SL, Clements M, Sparrow DB, Conlon R, Beddington RSP. Axial skeletal defects caused by mutation in the spondylocostal dysplasia/pudgy gene Dll3 are associated with disruption of the segmentation clock within the presomitic mesoderm. Development. 2002;129:1795–806.
50. Bulman MP, Kusumi K, Frayling TM, McKeown C, Garrett C, Lander ES, et al. Mutations in the human Delta homologue, DLL3, cause axial skeletal defects in spondylocostal dysostosis. Nature Genet. 2000;24:438–41.
51. Bonafé L, Giunta C, Gassner M, Steinmann B, Superti-Furga A. A cluster of autosomal recessive spondylocostal dysostosis caused by three newly identified DLL3 mutations segregating in a small village. Clin Genet. 2003;64:28–35.
52. Turnpenny PD, Whittock N, Duncan J, Dunwoodie S, Kusumi K, Ellard S. Novel mutations in DLL3, a somitogenesis gene encoding a ligand for the notch signaling pathway, cause a consistent pattern of abnormal vertebral segmentation in spondylocostal dysostosis. J Med Genet. 2003;40:333–9.
53. Whittock NV, Sparrow DB, Wouters MA, Sillence D, Ellard S, Dunwoodie SL, et al. Mutated MESP2 causes spondylocostal dysostosis in humans. Am J Hum Genet. 2004;74:1249–54.
54. Cornier AS, Staehling-Hampton K, Delventhal KM, Saga Y, Caubet JF, Sasaki N, et al. Mutations in the MESP2 gene cause spondylothoracic dysostosis/Jarcho-Levin syndrome. Am J Hum Genet. 2008;82:1334–41.
55. Cornier AS, Ramírez N, Arroyo S, Acevedo J, García L, Carlo S, et al. Phenotype characterisation and natural history of spondylothoracic dysplasia syndrome: a series of 27 new cases. Am J Med Genet. 2004;128A:120–6.
56. Sparrow DB, Chapman G, Wouters MA, Whittock NV, Ellard S, Fatkin D, et al. Mutation of the LUNATIC FRINGE gene in humans causes spondylocostal dysostosis with a severe vertebral phenotype. Am J Hum Genet. 2006;78:28–37.
57. Haines N, Irvine KD. Glycosylation regulates notch signaling. Nat Rev Mol Cell Biol. 2003;4:786–97.
58. Evrard YA, Lun Y, Aulehla A, Gan L, Johnson RL. Lunatic fringe is an essential mediator of somite segmentation and patterning. Nature. 1998;394:377–81.
59. Zhang N, Gridley T. Defects in somite formation in lunatic fringe-deficient mice. Nature. 1998;394:374–7.
60. Sparrow DB, Guillén-Navarro E, Fatkin D, Dunwoodie SL. Mutation of Hairy-and-Enhancer-of-Split-7 in humans causes spondylocostal dysostosis. Hum Mol Genet. 2008;17:3761–6.

61. Sparrow DB, Sillence D, Wouters MA, Turnpenny PD, Dunwoodie SL. Two novel missense mutations in HAIRY-AND-ENHANCER-OF-SPLIT-7 in a family with spondylocostal dysostosis. Eur J Hum Genet. 2010;18:674–9.
62. Kageyama R, Niwa Y, Isomura A, González A, Harima Y. Oscillatory gene expression and somitogenesis. Wiley Interdiscip Rev Dev Biol. 2012;1:629–41.
63. Bessho Y, Miyoshi G, Sakata R, Kageyama R. Hes7: a bHLH-type repressor gene regulated by notch and expressed in the presomitic mesoderm. Genes Cells. 2001;6:175–85.
64. Bessho Y, Sakata R, Komatsu S, Shiota K, Yamada S, Kageyama R. Dynamic expression and essential functions of Hes7 in somite segmentation. Genes Dev. 2001;15:2642–7.
65. Sparrow DB, Faqeih EA, Sallout B, Alswaid A, Ababneh F, Al-Sayed M, et al. Mutation of HES7 in a large extended family with spondylocostal dysostosis and dextrocardia with situs inversus. Am J Med Genet. 2013;161A(9):2244–9.
66. Sparrow DB, McInerney-Leo A, Gucev ZS, Gardiner B, Marshall M, Leo PJ, et al. Autosomal dominant spondylocostal dysostosis is caused by mutation in TBX6. Hum Mol Genet. 2013;22(8):1625–31.
67. Gucev ZS, Tasic V, Pop-Jordanova N, Sparrow DB, Dunwoodie SL, Ellard S, et al. Autosomal dominant spondylocostal dysostosis in three generations of a Macedonian family: negative mutation analysis of DLL3, MESP2, HES7, and LFNG. Am J Med Genet. 2010;152A:1378–82.
68. Yasuhiko Y, Haraguchi S, Kitajima S, Takahashi Y, Kanno J, Saga Y. Tbx6-mediated notch signaling controls somite-specific Mesp2 expression. Proc Natl Acad Sci U S A. 2006;103:3651–6.
69. McInerney-Leo AM, Sparrow DB, Harris JE, Gardiner BB, Marshall MS, et al. Compound heterozygous mutations in RIPPLY2 associated with vertebral segmentation defects. Hum Mol Genet. 2015;24:1234–42.
70. Wu N, Ming X, Xiao J, Wu Z, Chen X, Shinawi M, et al. TBX6 null variants and a common hypomorphic allele in congenital scoliosis. N Engl J Med. 2015;372:341–50.
71. Lefebvre M, Duffourd Y, Jouan T, Poe C, Jean-Marçais N, Verloes A, et al. Autosomal recessive variations of TBX6, from congenital scoliosis to spondylocostal dysostosis. Clin Genet. 2017;91:908–12.
72. Sandbacka M, Laivuori H, Freitas É, Halttunen M, Jokimaa V, Morin-Papunen L, et al. TBX6, LHX1 and copy number variations in the complex genetics of Müllerian aplasia. Orphanet J Rare Dis. 2013;8:125.
73. Aymé S, Preus M. Spondylocostal/spondylothoracic dysostosis: the clinical basis for prognosticating and genetic counselling. Am J Med Genet. 1986;24:599–606.
74. Roberts AP, Conner AN, Tolmie JL, Connor JM. Spondylothoracic and spondylocostal dysostosis. J Bone Jt Surg. 1988;70B:123–6.
75. Perez-Comas A, Garcia-Castro JM. Occipito-facial-cervicothoracic-abdomino-digital dysplasia: Jarcho Levin syndrome of vertebral anomalies. J Pediatr. 1974;85:388–91.
76. Karnes PS, Day D, Berry SA, Pierpont ME. Jarcho-Levin syndrome: four new cases and classification of subtypes. Am J Med Genet. 1991;40(3):264–70.
77. Martínez-Frías ML, Urioste M. Segmentation anomalies of the vertebras and ribs: a developmental field defect: epidemiologic evidence. Am J Med Genet. 1994;49:36–44.
78. Rastogi D, Rosenzweing EB, Koumbourlis A. Pulmonary hypertension in Jarcho Levin syndrome. Am J Med Genet. 2002;107:250–2.
79. Bannykh SI, Emery SC, Gerber J-K, Jones KL, Benirschke K, Masliah E. Aberrant Pax1 and Pax9 expression in Jarcho Levin syndrome: report of 2 caucasian siblings and literature review. Am J Med Genet. 2003;120A:241–6.
80. Cornier AS, Ramirez N, Carlo S, Reiss A. Controversies surrounding Jarcho-Levin syndrome. Current Opinion Pediatr. 2003;15:614–20.
81. Cantú JM, Urrusti J, Rosales G, Rojas A. Evidence for autosomal recessive inheritance of costovertebral dysplasia. Clin Genet. 1971;2:149–54.
82. Bartsocas CS, Kiossoglou KA, Papas CV, Xanthou-Tsingoglou M, Anagnostakis DE, Daskalopoulou HD. Costovertebral dysplasia. Birth Defects OAS. 1974;X(9):221–6.
83. David TJ, Glass A. Hereditary costovertebral dysplasia with malignant cerebral tumour. J Med Genet. 1983;20:441–4.

84. Rimoin DL, Fletcher BD, McKusick VA. Spondylocostal dysplasia. Am J Med Genet. 1968;45:948–53.
85. Silengo MC, Cavallaro S, Francheschini P. Recessive spondylocostal dysostosis: two new cases. Clin Genet. 1978;13:289–94.
86. Moseley JE, Bonforte RJ. Spondylothoracic dysplasia – a syndrome of congenital anomalies. Am J Roentgenol. 1969;106:166–9.
87. Pochaczevsky R, Ratner H, Perles D, Kassner G, Naysan P. Spondylothoracic dysplasia. Radiology. 1971;98:53–8.
88. Solomon L, Jimenez B, Reiner L. Spondylothoracic dysostosis. Arch Pathol Lab Med. 1978;102:201–5.
89. Kozlowski K. Spondylo-costal dysplasia. Fortschr Röntgenstr. 1984;140:204–9.
90. Ohashi H, Sugio Y, Kajii T. Spondylocostal dysostosis: report of three patients. Jpn J Hum Genet. 1987;32:299–303.
91. Jarcho S, Levin PM. Hereditary malformation of the vertebral bodies. Bull Johns Hopkins Hosp. 1938;62:216–26.
92. Berdon WE, Lampl BS, Cornier AS, Ramirez N, Turnpenny PD, Vitale MG, et al. Clinical and radiological distinction between spondylothoracic dysostosis (Lavy-Moseley syndrome) and spondylocostal dysostosis (Jarcho-Levin syndrome). Pediatr Radiol. 2011;41(3):384–8.
93. Casamassima AC, Morton CC, Nance WE, Kodroff M, Caldwell R, Kelly T, et al. Spondylocostal dysostosis associated with anal and urogenital anomalies in a Mennonite sibship. Am J Med Genet. 1981;8:117–27.
94. Daikha-Dahmane F, Huten Y, Morvan J, Szpiro-Tapia S, Nessmann C, Eydoux P. Fetus with Casamassima-Morton-Nance syndrome and an inherited (6;9) balanced translocation. Am J Med Genet. 1998;80:514–7.
95. Poor MA, Alberti O Jr, Griscom NT, Driscoll SG, Holmes LB. Nonskeletal malformations in one of three siblings with Jarcho-Levin syndrome of vertebral anomalies. J Pediatr. 1983;103:270–2.
96. Offiah A, Alman B, Cornier AS, Giampietro PF, Tassy O, Wade A, et al. Pilot assessment of a radiologic classification system for segmentation defects of the vertebrae. Am J Med Genet. 2010;152A:1357–71.

Chapter 4
The Genetics Contributing to Disorders Involving Congenital Scoliosis

Nan Wu, Philip Giampietro, and Kazuki Takeda

Introduction

Congenital scoliosis (CS) is a congenital deformity of the spine. The spine is derived from somites by a process called somitogenesis, characterized by intricate interactions of genes, signaling pathways, and related effectors. CS can represent an isolated malformation or part of a syndrome with other clinical features such as renal, cardiac, gastrointestinal, and limb malformations. Genetic factors are involved in the development of CS, especially perturbations of genes in FGF, WNT, Notch, and TGFβ signaling pathways. Analysis of the current understanding of these conditions and associated genetic mechanisms will be employed to promote insight into the genetic susceptibilities contributing to CS. Genetic approaches utilized to understand the etiology of CS will also be discussed.

N. Wu (✉)
Department of Orthopedic Surgery, Peking Union Medical College Hospital, Peking Union Medical College and Chinese Academy of Medical Sciences, Beijing, China

Beijing Key Laboratory for Genetic Research of Skeletal Deformity, Beijing, China

Medical Research Center of Orthopedics, Chinese Academy of Medical Sciences, Beijing, China
e-mail: dr.wunan@pumch.cn

P. Giampietro
Department of Pediatrics, Drexel University College of Medicine, Philadelphia, PA, USA

K. Takeda
Laboratory of Bone and Joint Diseases, Center for Integrative Medical Sciences, RIKEN, Tokyo, Japan

Department of Orthopedic Surgery, Keio University School of Medicine, Tokyo, Japan

K. Kusumi, S. L. Dunwoodie (eds.), *The Genetics and Development of Scoliosis*, https://doi.org/10.1007/978-3-319-90149-7_4

Background

CS is a congenital deformity of the spine with a prevalence of 0.5–1% in the general population [1, 2]. It is characterized by the presence of vertebral malformations (VMs) which result in a rapid progression of spinal curvature and multiple comorbidities. For example, pulmonary function is often compromised, and paralysis may occur in some instances. The spinal deformity may significantly impact the physical and the psychological health of affected patients. About 50% of CS is associated with deformities of other organs [3–5], including renal, cardiac, and intraspinal defects [6], or is a part of an underlying genetic syndrome. In the absence of effective treatment, patients suffer from progression of their spine deformity [7] and significantly increased mortality rate [8]. The surgical cost of managing CS is estimated to be $150,000 per operation with frequent follow-up examinations or operations, which all add to the financial burden for the family and the society [9]. Previous studies have demonstrated that both genetic and environmental factors are involved in the development of CS [10, 11].

Embryological Basis of CS

In the past few decades, research in spine development has unraveled possible etiologies of CS. Accumulating evidence supports a relationship between genotype and phenotype in CS patients with an embryological basis. In human and other mammals, vertebral bodies are derived from somites through a process called somitogenesis, which represents a harmonic convergence of multiple signaling pathways, related genes, and related effectors. Mutations in genes associated with somitogenesis or disruption of the symmetric gene modulation may eventually contribute to the occurrence of CS.

In vertebrate embryogenesis, the paraxial mesoderm originates from progenitors initially located in a superficial layer of the embryo (epiblast), which are internalized later in gastrulation and form the presomitic mesoderm (PSM). Subsequently, the paraxial mesoderm undergoes segmentation and lies on the lateral sides of the neural tube. The primitive streak differentiates into a mass of cells called the tail bud. The tail bud is located at the posterior tip of the embryo, which contains progenitors of the PSM that contribute to subsequent tissue formation. In this process, the structure of the somite is gradually formed through a synchronous and rhythmic fashion. The somites give rise to the vertebrae, muscles, tendons, and ligaments of the spine.

Embryonic development of somites from PSM is regulated by a variety of factors. Detailed mechanisms underlying the networked interactions of these factors have been illustrated by multiple models, with the clock-wavefront model widely accepted [12]. In the clock-wavefront model, the PSM is progressively segmented into repetitive somites, driven by the periodic activation of Notch, WNT, and FGF signaling pathways [13]. Segmentation clock and the wavefront model is composed of the sequential interactions of a variety of signals. The segmentation clock is initiated by the oscillatory expression of the *HAIRY1* gene and a variety of other genes called oscillators, with several genes expressed periodically representing the "clock" portion of the model [14, 15]. The wavefront portion of the model corresponds to FGF and WNT signal gradient [16]. In the paraxial mesoderm, the formation of segmentation clock depends on both the periodic expression of related genes and their repression due to negative feedback mechanisms [17]. Periodic signals of clock and wavefront enable the PSM to form somites individually in a single periodic cycle.

In somitogenesis, segmentation occurs after the formation of somites, when the formed somites receive a clock signal. For example, *MESP2* is activated by NICD (Notch signaling) and *TBX6*. *MESP2* are initially expressed in a restricted area (one segment length), and subsequently *RIPPLY1/2* is expressed in the posterior half region, thus defining future segment boundaries according to the area of the signal [18]. The downstream target gene *RIPPLY2* is activated, which is thought to be a negative feedback inhibitor of *MESP2* and *TBX6*. The process contributes to the definition of the anterior boundary of the newly formed segment [19]. Inactivation of *MESP1* and *MESP2* results in failure of paraxial mesoderm formation, thus indicating their importance in somitogenesis [20, 21].

One of the most notable features of somitogenesis is bilateral symmetry. Segmentation clocks are synchronized on both sides, ensuring the symmetric formation of somites in both sides of the PSM. Retinoic acid (RA), a derivative of vitamin A, has been implicated as a key regulator of the symmetry in somitogenesis [22–24]. The process of somitogenesis depends upon close interaction between the clock and wavefront and genes which are essential for neural development to form the normal body axis structures including somites and the neural tube.

In summary, the process of somitogenesis is controlled by Notch, WNT, and FGF signaling-related genes. They coordinate the key steps during somitogenesis including segmentation, bilateral symmetry, and vertebral formation. Disruption of the signaling pathways and malfunction of regulators in these processes are of importance in the pathogenesis in CS. Genes associated with somitogenesis and vertebral development are summarized in Table 4.1.

Approaches in Understanding the Genetic Etiology of CS

Vertebral development occurs through a synchronous convergence of multiple pathways and several dozen genes. As described before, genes in the FGF, WNT, and Notch signaling pathways are liable to mutate in CS patients. Previous work reported several candidate loci by numerous methods.

Cytogenetic analysis demonstrated the presence of a paracentric inversion inv. (q22.2q23.3) [8] segregating in those individuals with phenotypic manifestations of Klippel-Feil syndrome (KFS) defined by the presence of a short neck, low posterior

Table 4.1 Somitogenesis-associated genes

Category	Gene name[a]
NOTCH related [15, 25]	*HEY2*, *PSEN1*, *HES1*, *DLL1*, ***JAG1***, ***LFNG***, ***DLL3***, ***MESP1***, ***MESP2***, *NKD1*, *RTF*, ***RIPPLY***
FGF related [15]	*DUSP6*, *FRS2*, *GRB2*, *SOS1*, *FGF8*, *BCL2L11*, *EFNA1*, *EPHB2*, *HSPG2*, *SHH*, *PTPN11*, *GAB1*
WNT related [15]	***WNT3A***, *FZD7*, *FZD5*, *CDC73*, *PHLDA1*, *DVL2*, *HDAC9*, *DACT1*, *TNFRSF1*, *FZD3*, *SPRY2*, *FZD6*, *HAS2*, *MYC*, *APC*, *SMAD4*, *LRP5*, *DKK1*, *FRZB*, *TCF15*, *FZD1*, *CTBP1*, *FZD9*, *SMARCA5*, *CER1*
HOX related [26]	*HOXC8*, *HOXC4*, *HOXD11*, *HOXD10*, *HOXD3*, *HOXA7*, *HOXB7*, *PAF1*
PAX [27]	*PAX9*, *PAX1*, *PAX7*
Other [28–32]	*RAB23*, *IHH*, *PLXDC1*, *TWIST1*, *GLI3*, *FLNB*, *SLC35A3*, *MXD4*, *PDFGFRA*, ***TBX6***, *ACD*, *MID1*, ***GDF3***, ***GDF6***, ***POLR1D***, *COL8A1*, ***T (brachyury)***, ***MEOX1***

[a]**Bold** denotes mutated gene associated with VMs; mutations have been identified in patterning genes outside known somitogenic pathways in "Other" category. Mutations have not been identified in the majority of somitogenesis associated genes identified in other species

hairline, cervical vertebral fusion, and limitation of range of motion of the neck [28]. This information was subsequently used to identify mutations in *GDF6* in both sporadic and familial cases of KFS. Additional cytogenetic evidence for loci contributing to the development of CS includes the observation of additional chromosomal rearrangements, including de novo balanced reciprocal translocation – t(5;17) (q11.2;q23) [33], de novo pericentric inversion inv. [2] (p12q34) [34], and translocation – t(5;8) (q35.1;p21.1) [35]. Traditional linkage methodologies utilizing single nucleotide polymorphisms (SNP) or short tandem repeat (STR) polymorphisms may be applied to large families with CS when cytogenetic etiologies have been eliminated.

However, traditional linkage analysis has significant limitations to identify causative genes since isolated CS often represents a sporadic occurrence. Candidate gene analyses offered a reasonable alternative method for studies. Utilizing mouse-human synteny analysis, 27 eligible loci, 21 of which cause VMs in mouse, have been identified [36, 37]. In a phenotypically well-defined cohort of patients with CS, five candidate genes extrapolated from murine models were chosen for analysis [27, 38–41]. The VMs represented among this cohort spanned the entire length of the spine and are described in greater detail in a prior communication [27]. Sequence variants in *PAX1*, *DLL3*, *WNT3A*, and *T (Brachyury)* associated with decreased penetrance have been identified in patients with VMs and were seen at low frequency or not detected in healthy controls [38]. In a cohort of 79 CS cases, a screen for variants in *DLL3*, *MESP2*, and *HES7* (genes associated with causing severe VMs such as spondylocostal dysostosis; see below) was conducted. One family carried a variant in *MESP2* and another in *HES7*. In both families, the penetrance and expressivity of the mutation were variable; importantly the variants were shown to impair the in vitro function of the transcription factors encoded by these genes [11]. In another cohort containing 154 CS patients, genetic polymorphisms of *LMX1A* were found to be associated with susceptibility to CS in Chinese Han population [42].

Biallelic germline mutations in *CDK10* (cyclin-dependent kinase 10) have been identified in five consanguineous Saudi Arabian families with growth retardation, vertebrae fusion or hemivertebrae, and developmental delay. CDK10 is a protein kinase which has a regulatory role in transcription and cilia growth [43]. *Cdk10* knockout mice showed several bone defects affecting axial skeleton and longer cilia, suggesting that ciliary defects might contribute to the phenotype of individuals with *CDK10* mutations [43].

A novel missense mutation has been identified in *SLC35A3*, a gene associated with complex vertebral malformation in cattle [44, 45]. The patient's features included butterfly and hemivertebrae distributed throughout the spine, cleft palate, patent foramen ovale, patent ductus arteriosus, posterior embryotoxon, shortened limbs, and facial dysmorphism [46].

Chromosome microarray (previously referred to as array CGH) is a technique that was developed in order to determine alterations in dosage distribution of small DNA segments throughout the entire genome. An advantage of chromosome microarray is the ability to identify potential regions of micro-aneuploidy associated with VMs across the entire genome, instead of limited detection of only one focused region. Because etiology of CS is heterogeneous and may involve multiple genetic defects, many of which remain to be identified, this approach represents an efficient screening tool for defining additional loci that may harbor genetic defects underlying etiology of these disorders, although it may lack power to identify a single candidate gene. Chromosome microarray has successfully identified some potential regions associated with CS, with the identification of a common deletion encompassing the *TBX6* gene, identifying a major advance in our understanding of the causes of CS (see below) [29].

TBX6 Variants Contributing to CS

In clinical practice, a proportion of patients with CS do not display notably additional organic deformities. These cases are of great interest in the genomic research because they represent a phenotypically distinct group to study the pathogenesis of VMs.

Variations of genes involved in somitogenesis have been studied a decade before. Ghebranious et al. [38] sequenced a panel of patients with heterogeneous types of CS by a panel of genes associated with signal pathways in somitogenesis including *PAX1*, *DLL3*, *SLC35A3*, *WNT3A*, *TBX6*, and *T(Brachyury)*. They sequenced the complete *T(Brachyury)* and the coding regions, splice junctions, and 500 bp of the promoter region in *TBX6*. Three unrelated patients harbored the same c.1013C > T transition in exon 8 of the *T* gene, and no *TBX6* sequence variation was identified. Fei et al. [47] genotyped two known SNPs in *TBX6* among 254 Chinese Han subjects (127 CS patients and 127 controls). For the single SNP analysis, the allele frequencies of rs2289292 (SNP1, chr16:30005131, G/A, exon 8) and rs3809624 (SNP2, chr16:30010303, A/G, 5′-untranslated region) were significantly different between cases and controls ($P = 0.017$ and 0.033, respectively). The haplotype analysis showed a significant association between SNP1/SNP2 and CS cases ($P = 0.017$), with the G-A haplotype more frequently observed in controls (odds ratio, 0.71; 95% confidence interval, 0.51–0.99).

The *TBX6* gene is known as *T-box 6*, a member of the T-box family, and encodes a transcription factor which plays an important role in the regulation of development processes [48]. *TBX6* is localized to 16p11.2, with a 6091 bp in size, and contains 9 exons according to updated information. It has been reported that the interaction between *TBX6* and genes involved in clock-wavefront model or *TBX6* itself will result in abnormal formation of somites, contributing to CS [49–53].

Several reports showed that the copy number variations (CNVs) in the 16p11.2 region might be associated with CS phenotype. Shimojima et al. [54] reported a 3-year-old boy with developmental delay; inguinal hernia; hemivertebrae of T10, T12, and L3; a missing right 12th rib; and hypoplasia of the left 12th rib. The patient had a 593-kb interstitial deletion of 16p11.2, and the mother had the same deletion identified by chromosome microarray. Al-Kateb et al. [55] analyzed radiologic data obtained from ten patients with 16p11.2 CNV (nine with deletions and one with duplication). Eight of them had CS, and the remaining two had idiopathic scoliosis (IS). They additionally reviewed five patients reported previously with 16p11.2 rearrangement and similar skeletal abnormalities, concluding that two of them were affected with CS while the others had IS. Although many studies reported the association between 16p11.2 CNV and CS, the exact mechanism was still unclear at that time. Subsequently, Wu et al. [29] elaborated that *TBX6* null variants and a *TBX6* common hypomorphic allele together contribute to CS in a compound inheritance model. In a group of 161 Han Chinese patients with sporadic unrelated CS, CNV analysis identified 17 heterozygous *TBX6* null mutations in those persons affected with CS. This included 12 instances of a recurrent 16p11.2 deletion affecting *TBX6* and 5 single nucleotide variants (1 nonsense and 4 frameshift mutations). No *TBX6* mutations were identified in the control group. The identification of phenotypically normal individuals with 16p11.2 microdeletions and the discordant intrafamilial phenotypes of CS in 16p11.2 microdeletion carriers suggested that heterozygous *TBX6* null mutations are insufficient to cause CS. Notably, they went on to identify another common (about 44% in Asian descent and 33% in European descent) haplotype which was showed as a hypomorphic allele *in trans* with the *TBX6* null mutations. This Compound Inheritance of both Rare and Common (CIRC) model accounted for up to 11% of sporadic CS cases. These findings were validated in an additional cohort and worldwide multicenter case series of 16p11.2 microdeletions and further replicated by the following studies in a Japanese CS cohort [56] and a French SDV (segmentation defects of the vertebra) cohort [57].

Syndromes and Disorders Associated with VMs

According to the classification scheme of International Consortium for Vertebral Anomalies and Scoliosis (ICVAS), several types of CS can be classified as syndromes [58]. Patients with those syndromes share VM and are different from each other in associated systemic phenotypes. Different genes and signaling pathways could be involved in the pathogenesis of these syndromes, compared with the cases studied above. These mechanisms should be more widely involved in the development of multiple organs or systems affected in these patients.

Spondylocostal Dysostosis

Spondylocostal dysostosis (SCD) is a heterogeneous group of spinal disorders characterized by multiple segmentation defects of the vertebrate, malalignment of the ribs, and often reduction in rib numbers. Distinct phenotypic subtypes, each with a distinctive genetic defect, are classified (Table 4.2). Limited by existing genetic technology at the time, prior studies only focused on several candidate genes. Due to these limitations, the process to identify SCD-associated candidate genes was a lengthy one. With the development of next-generation sequencing (NGS) and decrease of sequencing costs, it is possible to cast a wider net and test for mutations in all possible genes related to vertebral development in a single test.

Up to now, six genes have been demonstrated to be involved in the pathogenesis of SCD (Table 4.2) [81]. SCD will be discussed in detail in Chap. 3: however, some discussion with respect to mutation of *RIPPLY2* is warranted here. McInerney-Leo et al. [25] reported two VM patients within one family; they identified compound heterozygous mutations of *RIPPLY2* gene (a novel truncating variant c.A238T: p.Arg80* and a low frequency one c.240-4 T > G) segregating with the phenotype from whole exome sequence (WES). The *RIPPLY2* gene was classified as causing SCD type 6 in OMIM (Online Mendelian Inheritance in Man). However, the phenotype of patients in their report was hemivertebrae, which does not seem like a classical manifestation of SCD. These phenomena suggest that *RIPPLY2* might have some overlap in genetic etiology with other genes which correlate with vertebral malformation, such as *TBX6*.

Klippel-Feil Syndrome

KFS, also known as Klippel-Feil anomaly, has been identified as early as 500 B.C. in an Egyptian mummy and first arose by Klippel and Feil's description of a French tailor with massive cervical fusion [82]. The hallmark of KFS is the presence of cervical vertebral fusion abnormalities. Associated features observed in less than 50% of affected individuals include short neck, low posterior hairline, and limited neck motion. Additional multisystem malformations, including neural tube defects;

Table 4.2 Syndromic CS-related genes

Disease	MIM#	Phenotype	Gene	Reference
Spondylocostal dysostosis	277300	Multiple vertebral dysgenesis in the thoracic region with multiple misaligned ribs	*DLL3*	[59]
	605195	Multiple hemivertebrae in thoracic region, with lumbar vertebrate more angular and irregular	*MESP2*	[60]
	609813	Multiple vertebral malformation in cervical, thoracic, and lumbar spine	*LFNG*	[61]
	613686	Multiple and contiguous vertebral segmentation defects involving all spinal regions, but mainly the thoracic region	*HES7*	[62]
	122600	Mixture of hemivertebrae and vertebral blocks. Ribs are relatively mild affected	*TBX6*	[63]
	616566	Deficiency of posterior element of cervical, vertebral malformation in cervical and thoracic region	*RIPPLY2*	[25]
Klippel-Feil syndrome	118100	Defects in the formation or segmentation of the cervical vertebrae, resulting in a fused appearance. The clinical triad consists of short neck, low posterior hairline, and limited neck movement	*GDF6*	[64]
	613702		*GDF3*	[30]
	214300		*MEOX1*	[31, 65]
	616549		*MYO18B*	[66]
	–		*PAX1*	[67]
	–		*RIPPLY2*	[68]
	–		*POLRID*	[32]
Alagille syndrome	118450	Abnormal vertebrae (“butterfly” vertebrae) and decrease in interpedicular distance in the lumbar spine; neonatal jaundice; posterior embryotoxon and retinal pigmentary changes; pulmonic valvular stenosis as well as peripheral arterial stenosis; absent deep tendon reflexes and poor school performance; broad forehead, pointed mandible, and bulbous tip of the nose and in the fingers, varying degrees of foreshortening	*JAG1* *NOTCH2*	[69, 70]
VACTERL	314390	Vertebral anomalies, anal atresia, cardiac malformations, tracheoesophageal fistula, renal anomalies, and limb anomalies	*ZIC3*	[71–73]
			TRAP1	[74]
			PCSK5	[75]
			FOXF1	[76]
Congenital NAD deficiency disorder 1	617660	Vertebral segmentation abnormalities, congenital cardiac defects, renal defects, and distal mild limb defects. Additional features are variable	*HAAO*	[77]
Congenital NAD deficiency disorder 2	617661	Vertebral segmentation abnormalities, congenital cardiac defects, renal defects, and distal mild limb defects. Additional features are variable	*KYNU*	[77]

(continued)

Table 4.2 (continued)

Disease	MIM#	Phenotype	Gene	Reference
Facio-auriculo-vertebral spectrum	164210	Unilateral deformity of the external ear and small ipsilateral half of the face with epibulbar dermoid and vertebral anomalies	*SALL1*	[78]
			BAPX1	[79]
			MYT1	[80]

ENT (ear, nose, and throat) defects; thoracic cage abnormalities; pulmonary, cardiovascular, and other skeletal anomalies; genitourinary abnormalities; myopathy; neuropathy; and cognitive disorders, may occur in association with KFS [83, 84]. While the majority of the cases represent sporadic occurrences within a particular family, both autosomal dominant and autosomal recessive forms of KFS have been reported. To date, four disease-causing candidate genes and two disease-associated genes have been identified (Table 4.2). *GDF6* and *GDF3* genes have been reported causing autosomal dominant KFS. A pericentric inversion inv. [8] (q22.2q22.3) segregating in a four-generation family with KFS [64] has led to the identification of mutations in *GDF6*, a member of the BMP family, in familial and sporadic cases of KFS. In a three-generation family with autosomal dominant KFS, a missense mutation c.746C > A (p.A249E) was identified. In addition, a recurrent mutation in a highly conserved residue c.866 T > C (p.L289P) of *GDF6* was identified in 2 of 121 sporadic cases of KFS [28]. Ye et al. [30] described a three-generation North American family with a clinical spectrum of ocular and skeletal phenotypes. Two patients had cervical fusion at C5–6 and C3–4. A heterozygous mutation p.R266C in *GDF3* was identified that segregated with the skeletal phenotype in four family members. Morpholino knockdown of *dvr1*, the *GDF3*/*GDF1* zebrafish homolog, recapitulated ocular and skeletal phenotypes seen in humans.

The occurrence of KFS in sibs and in consanguineous families suggests autosomal recessive inheritance of the disorder. Homozygous *MEOX1* [31, 65] and *MYO18B* [85] truncating mutations have also been identified in different consanguineous families. A c.670G > A (p.Q84*) mutation in the *MEOX1* gene was identified in affected family members with KFS in a Turkish consanguineous marriage family. Two additional consanguineous families with members affected with KFS were found to harbor mutations in *MEOX1*. MEOX1 plays an important, nonredundant role in maintaining sclerotome polarity and the formation of craniocervical joints, and *Meox1* null mutant mice have defects in the axial skeleton [86]. *Meox1* maintains the appropriate expression of downstream genes such as *Bapx1*, *Tbx18*, and *Uncx*, which regulate somite development [86, 87]. *MYO18B* is another locus which accounts for the occurrence of KFS. Two patients from unrelated family shared similar phenotype characterized by KFS and myopathy. The compound phenotypes were found to be associated with a null mutation (c.6905C > A, p.S2302*) in *MYO18B*. MYO18B functions as an ancillary protein to myosin 2 and is highly expressed in both somites and striated muscles in mouse [88].

There is evidence that mutations in the *PAX1* gene also contribute to the occurrence of KFS [67]. *Pax1* is expressed in the sclerotome, which gives rise to the vertebrae and ribs [89, 90]. Medial sclerotome condensation fails to occur at the lumbosacral level in the *undulated* mouse, thus preventing the formation of intervertebral disks and vertebral bodies [91]. Mutations in *PAX1* were identified in 3 of 63 patients diagnosed with KFS. One of these patients and the asymptomatic mother of the patient carried a (c.224C > G, p.P61A) missense mutation 38 bp upstream from the paired-box region, which is associated with DNA binding of the PAX1 protein. Assuming this mode of inheritance is autosomal dominant, reduced penetrance could be postulated if this mutation promotes causation (Table 4.2).

Karaca et al. [68] reported a homozygous *RIPPLY2* frameshift mutation (c.299delT, p.L100 fs) in a KFS together with heterotaxy pedigree. Considering the mechanism that *Ripply2* is a direct transcriptional target of *Tbx6* and *Mesp2*, they proposed that *RIPPLY2* is a novel gene for autosomal recessive KFS (Table 4.2).

Whole exome sequence analysis of an affected father and daughters with phenotypic features consistent with KFS including cervical segmentation defects, cleft palate, Sprengel deformity, and sensorineural hearing loss (suggestive of Treacher Collins syndrome) revealed a *POLR1D* mutation (c.T332C, p.L111P) [32]. Haploinsufficiency of *POLR1D* has previously been reported in association with altered ribosomal levels supporting development of neural crest-derived craniofacial structures. This finding supports the extension of the genotypic spectrum of VMs to mutations in genes outside of the FGF, Notch, and WNT signaling pathways (Table 4.2).

Syndromes Containing CS Phenotype

Alagille Syndrome

Alagille syndrome (ALGS) is a complex multisystem disorder characterized by the presence of three out of seven major clinical criteria which include cholestasis with bile duct paucity on liver biopsy, congenital cardiac defects (with particular involvement of the pulmonary arteries), posterior embryotoxon in the eyes, characteristic facial features (broad forehead, deep set eyes, pointed chin), vertebral anomaly, and abnormalities of the kidneys and vasculature (often in the head and neck) [92]. The most common radiographic finding, observed with a frequency from 33% to 93%, is butterfly-shaped thoracic vertebrae, secondary to cleft abnormality of the vertebral bodies.

ALGS is inherited in an autosomal dominant manner with 50–70% of cases representing de novo occurrences. The rate of germline mosaicism may also be relatively high (about 8%) [93, 94]. Currently, two genes are associated with ALGS: *JAG1* located at 20p12.2 and *NOTCH2* located at 1p12 [69]. The majority of the cases (up to 95%) are caused by haploinsufficiency of the *JAG1* gene (encoding Jagged-1), either due to missense mutations (the majority) or deletions at the locus. To date, over 400 different *JAG1* gene mutations have been identified in ALGS patients (point mutations, microdeletions, and insertions), showing the variability in its mutagenesis and paucity of recurrent mutations [95]. A small percentage (<5%)

of ALGS are caused by mutations in *NOTCH2* [70, 92], which are more commonly associated with incompletely penetrant ALGS with renal disease [96].

JAG1 is a cell surface protein that functions as a ligand for one of four Notch transmembrane receptors, which are key signaling molecules in the Notch signaling pathway, an evolutionarily conserved pathway that is crucial in development. *NOTCH2* encodes a member of the Notch family transmembrane receptors. The Notch receptors (*NOTCH1*, *NOTCH2*, *NOTCH3*, and *NOTCH4* in humans) share structural characteristics, including an extracellular domain consisting of multiple epidermal growth factor-like repeats and an intracellular domain consisting of multiple, different domain types. The Notch family members play a role in a variety of developmental processes by controlling cell fate decisions.

No genotype-phenotype correlations exist between the clinical manifestations of ALGS and the specific *JAG1* and *NOTCH2* pathogenic variant types or the location of the mutation within the genes. A recent study using genome-wide association study identified a new locus (rs7382539) for ALGS that reached suggestive genome-level significance upstream of the thrombospondin 2 (*THBS2*) gene [97]. *THBS2* codes for an adhesive glycoprotein that mediates cell-cell and cell-matrix interactions associated with cell proliferation, apoptosis, and angiogenesis, as well as affecting Notch signaling. *THBS2* expression may further perturb *JAG1-NOTCH2* signaling in patients harboring a *JAG1* mutation and lead to a more severe phenotype with liver malformations, thus implicating *THBS2* as a plausible candidate genetic modifier of liver disease severity in ALGS (Table 4.2).

VACTERL Association

VACTERL association is a condition comprising multisystem congenital malformations, which is typically defined by the concurrence of at least three of the following component features: vertebral anomalies (V), anal atresia (A), cardiac malformations (C), tracheoesophageal fistula (TE), renal dysplasia (R), and limb abnormalities (L). The prevalence of VACTERL/VATER association is between 1/7000 and 1/40,000 [98]. Although the frequency of the six clinical features varies, vertebral anomalies are the most common observation in many cohorts of VACTERL association, which have been reported in approximately 60–95% of affected individuals [99].

VACTERL association is a rare and complex condition with highly heterogeneous etiology and manifestations. Although the clinical criteria for VACTERL association appear to be straightforward, the overlapping in either clinical manifestation or genetic finding is challenging for clinicians and geneticists. There is evidence for genetic factors contributing to VACTERL syndrome including single gene mutations, CNVs, structure variants, and mitochondrial dysfunction [100]. Among these genetic factors, several candidate gene mutations and CNVs have been reported to be related with different vertebrae phenotypic features. For example, different types of *ZIC3* variants, including point mutations, deletions, and polyalanine expansion, have been reported to be responsible for both VACTERL and VACTERL-like association [71, 72]. Mutations in *TRAP1* [74], *PCSK5* [75] and *FOXF1* [76] have also been reported in VACTERL patients with hemivertebrae or butterfly vertebrae, while *DLL3* mutation was reported in patients with block vertebrae [40]. 13q deletion [101] and 19p13.3 microdeletion [102] were also found in VACTERL patients, but not all involved patients harbor VMs. These may imply that other modification factors were involved which warrant further investigation. The detailed descriptions regarding the involvement of VMs in VACTERL association can be found in a systemic review from Chen et al. [100].

Congenital NAD Deficiency Disorders

Recently, nicotinamide adenine dinucleotide (NAD) deficiency was robustly reported in a group of patients with multiple congenital malformations similar to VACTERL. Before this report, many causes of isolated organ defects had been identified, such as heart defects and vertebral malformation. But the genetic causes of isolated cardiac or vertebral defects appear to have little relevance when these defects occur in combination. In the study, Shi et al. [77] enrolled patients with congenital vertebral and heart malformations from four families. Whole exome/genome sequence identified biallelic pathogenic *HAAO* and *KYNU* variants in these patients. In vitro assay proved that loss of *HAAO* or *KYNU* activity leads to increased plasma levels of metabolites upstream of these enzymes and reduced levels downstream, including NAD. Homozygous null mutant mice showed that all the *Haao*$^{-/-}$ and *Kynu*$^{-/-}$ embryos had multiple defects, including defects in vertebral segmentation, heart defects, small kidney, cleft palate, talipes, syndactyly, and caudal agenesis, recapitulating the human phenotype. Importantly, these mice could be rescued by niacin (vitamin B3) supplementation, indicating that loss of embryonic NAD leads to embryo defects and death. The rescue experiment also indicates that vitamin B3 supplementation might prevent miscarriage and malformation in humans.

Facio-Auriculo-Vertebral Spectrum

Facio-auriculo-vertebral spectrum (FAVS), also known as Goldenhar syndrome and hemifacial microsomia, is a rare birth defect associated with the abnormal development of the first branchial arch and second branchial arch. Typical symptoms are usually present on the affected side, involving incomplete development of the external ear, middle ear, nose, mandible, masticatory muscles, facial muscles, and other facial tissues [103]. Other clinical features may include VMs, including severe scoliosis, fused cervical vertebrae, and spina bifida occulta of multiple lumbar vertebrae [104]. Some cases also displayed several atypical findings, including imperforate anus and rib fusions [105].

Although there may be genetic components, which lead to certain family patterns, FAVS is considered to have multifactorial etiologies which remain largely unknown. *SALL1* gene mutation (c.826C > T, p.R276*) has been identified in affected individuals with FAVS [78], and a c.1256 T > A (p.L419*) mutation has been reported in two sisters who are clinically discordant for this condition [106]. Strong allelic expression imbalance of *BAPX1* was observed in patients, and epigenetic dysregulation could predispose to FAVS [79]. A de novo nonsense mutation (c.25C > T, p.9 > *) in *MYT1* is reported in a female case by whole exome sequence, so that *MYT1* is hypothesized to be a candidate gene for FAVS [80] (Table 4.2).

In an autosomal dominant multigeneration pedigree in which five individuals are affected with FAVS, a 1.3 Mb duplication of chromosome 14q22.3 was identified in all affected individuals. Utilization of human craniofacial disease network signatures, mouse expression data, and dosage sensitivity predictions implicated *OTX2* as a potential causal gene [107]. The proband had medulloblastoma, consistent with the observation that overexpression of *OTX2* is associated with the occurrence of medulloblastoma. In a cohort of 51 patients with clinical features of FAVS, ten patients were found to have CNV involving the 22q11.21 genomic region [108]. All ten patients had microduplications of this region with four of them having two or more nonoverlapping microduplications of the 22q11.21 genomic region. Despite the occurrence of 22q11.21 microduplications in clinically unaffected individuals, the 22q11.21 genomic region may contribute to the regulation of branchial arch-derived structures.

Perspective

Currently, the classification of CS is mainly based on clinical manifestations. Mutations of genes in key signaling pathways could lead to different CS syndromes with similar phenotypes, while the same syndrome could be separated into several subtypes according to genotyping thus providing evidence for phenotypic and genetic heterogeneity. Panel/target sequencing, whole exome sequence (WES), and whole genome sequence (WGS) are now applied in molecular diagnosis [109, 110]. It is anticipated that these methods will be increasingly utilized in genetic diagnoses and molecular classification in CS patients, opening a new era for novel gene discovery and clinical practice. As genomic variation is characterized in different races, larger-scale multicenter cohorts with multiethnic studies are in great need to identify universal or possible differences in genes associated with CS among different racial groups. Using GWAS strategy in conjunction with NGS [111] may help localize additional candidate loci. As the causes for CS and VM are heterogeneous, a combination of various techniques needs to be implemented to identify the genetic etiology. Functional studies and animal models are necessary to ascertain pathogenicity of identified variants.

References

1. Wynne-Davies R. Congenital vertebral anomalies: aetiology and relationship to spina bifida cystica. J Med Genet. 1975;12:280–8.
2. Brand MC. Examination of the newborn with congenital scoliosis: focus on the physical. Adv Neonatal Care. 2008;8:265–73. 274–5
3. Beals RK, Robbins JR, Rolfe B. Anomalies associated with vertebral malformations. Spine (Phila Pa 1976). 1993;18:1329–32.
4. Basu PS, Elsebaie H, Noordeen MH. Congenital spinal deformity: a comprehensive assessment at presentation. Spine (Phila Pa 1976). 2002;27:2255–9.
5. Shen J, Wang Z, Liu J, et al. Abnormalities associated with congenital scoliosis: a retrospective study of 226 chinese surgical cases. Spine (Phila Pa 1976). 2013;38:814–8.
6. Hensinger RN. Congenital scoliosis: etiology and associations. Spine (Phila Pa 1976). 2009;34:1745–50.
7. Marks DS, Qaimkhani SA. The natural history of congenital scoliosis and kyphosis. Spine (Phila Pa 1976). 2009;34:1751–5.
8. Cahill PJ, Samdani AF. Early-onset scoliosis. Orthopedics. 2012;35:1001–3.
9. Kamerlink JR, Quirno M, Auerbach JD, et al. Hospital cost analysis of adolescent idiopathic scoliosis correction surgery in 125 consecutive cases. J Bone Joint Surg Am. 2010;92:1097–104.
10. Giampietro PF, Raggio CL, Blank RD, et al. Clinical, genetic and environmental factors associated with congenital vertebral malformations. Mol Syndromol. 2013;4:94–105.
11. Sparrow DB, Chapman G, Smith AJ, et al. A mechanism for gene-environment interaction in the etiology of congenital scoliosis. Cell. 2012;149:295–306.
12. Baker RE, Schnell S, Maini PK. A clock and wavefront mechanism for somite formation. Dev Biol. 2006;293:116–26.
13. Dequeant ML, Pourquie O. Segmental patterning of the vertebrate embryonic axis. Nat Rev Genet. 2008;9:370–82.
14. Palmeirim I, Henrique D, Ish-Horowicz D, et al. Avian hairy gene expression identifies a molecular clock linked to vertebrate segmentation and somitogenesis. Cell. 1997;91:639–48.
15. Dequeant ML, Glynn E, Gaudenz K, et al. A complex oscillating network of signaling genes underlies the mouse segmentation clock. Science. 2006;314:1595–8.
16. Dubrulle J, McGrew MJ, Pourquie O. FGF signaling controls somite boundary position and regulates segmentation clock control of spatiotemporal Hox gene activation. Cell. 2001;106:219–32.
17. Hubaud A, Pourquie O. Signalling dynamics in vertebrate segmentation. Nat Rev Mol Cell Biol. 2014;15:709–21.
18. Takahashi Y, Koizumi K, Takagi A, et al. Mesp2 initiates somite segmentation through the Notch signalling pathway. Nat Genet. 2000;25:390–6.
19. Zhao W, Ajima R, Ninomiya Y, et al. Segmental border is defined by Ripply2-mediated Tbx6 repression independent of Mesp2. Dev Biol. 2015;400:105–17.
20. Kitajima S, Takagi A, Inoue T, et al. MesP1 and MesP2 are essential for the development of cardiac mesoderm. Development. 2000;127:3215–26.
21. Chen W, Liu J, Yuan D, et al. Progress and perspective of TBX6 gene in congenital vertebral malformations. Oncotarget. 2016;7:57430–41.
22. Vermot J, Gallego LJ, Fraulob V, et al. Retinoic acid controls the bilateral symmetry of somite formation in the mouse embryo. Science. 2005;308:563–6.
23. Vermot J, Pourquie O. Retinoic acid coordinates somitogenesis and left-right patterning in vertebrate embryos. Nature. 2005;435:215–20.
24. Kawakami Y, Raya A, Raya RM, et al. Retinoic acid signalling links left-right asymmetric patterning and bilaterally symmetric somitogenesis in the zebrafish embryo. Nature. 2005;435:165–71.

25. McInerney-Leo AM, Sparrow DB, Harris JE, et al. Compound heterozygous mutations in RIPPLY2 associated with vertebral segmentation defects. Hum Mol Genet. 2014;24:1234–42.
26. Dias AS, de Almeida I, Belmonte JM, et al. Somites without a clock. Science. 2014; 343:791–5.
27. Giampietro PF, Raggio CL, Reynolds CE, et al. An analysis of PAX1 in the development of vertebral malformations. Clin Genet. 2005;68:448–53.
28. Tassabehji M, Fang ZM, Hilton EN, et al. Mutations in GDF6 are associated with vertebral segmentation defects in Klippel-Feil syndrome. Hum Mutat. 2008;29:1017–27.
29. Wu N, Ming X, Xiao J, et al. TBX6 null variants and a common hypomorphic allele in congenital scoliosis. N Engl J Med. 2015;372:341–50.
30. Ye M, Berry-Wynne KM, Asai-Coakwell M, et al. Mutation of the bone morphogenetic protein GDF3 causes ocular and skeletal anomalies. Hum Mol Genet. 2010;19:287–98.
31. Bayrakli F, Guclu B, Yakicier C, et al. Mutation in MEOX1 gene causes a recessive Klippel-Feil syndrome subtype. BMC Genet. 2013;14:95.
32. Giampietro PF, Armstrong L, Stoddard A, et al. Whole exome sequencing identifies a POLRID mutation segregating in a father and two daughters with findings of Klippel-Feil and Treacher Collins syndromes. Am J Med Genet A. 2015;167A:95–102.
33. Fukushima Y, Ohashi H, Wakui K, et al. De novo apparently balanced reciprocal translocation between 5q11.2 and 17q23 associated with Klippel-Feil anomaly and type A1 brachydactyly. Am J Med Genet. 1995;57:447–9.
34. Papagrigorakis MJ, Synodinos PN, Daliouris CP, et al. De novo inv(2)(p12q34) associated with Klippel-Feil anomaly and hypodontia. Eur J Pediatr. 2003;162:594–7.
35. Goto M, Nishimura G, Nagai T, et al. Familial Klippel-Feil anomaly and t(5;8)(q35.1;p21.1) translocation. Am J Med Genet A. 2006;140:1013–5.
36. Giampietro PF, Blank RD, Raggio CL, et al. Congenital and idiopathic scoliosis: clinical and genetic aspects. Clin Med Res. 2003;1:125–36.
37. Giampietro PF, Raggio CL, Blank RD. Synteny-defined candidate genes for congenital and idiopathic scoliosis. Am J Med Genet. 1999;83:164–77.
38. Ghebranious N, Blank RD, Raggio CL, et al. A missense T (Brachyury) mutation contributes to vertebral malformations. J Bone Miner Res. 2008;23:1576–83.
39. Ghebranious N, Raggio CL, Blank RD, et al. Lack of evidence of WNT3A as a candidate gene for congenital vertebral malformations. Scoliosis. 2007;2:13.
40. Giampietro PF, Raggio CL, Reynolds C, et al. DLL3 as a candidate gene for vertebral malformations. Am J Med Genet A. 2006;140:2447–53.
41. Ghebranious N, Burmester JK, Glurich I, et al. Evaluation of SLC35A3 as a candidate gene for human vertebral malformations. Am J Med Genet A. 2006;140:1346–8.
42. Wu N, Yuan S, Liu J, et al. Association of LMX1A genetic polymorphisms with susceptibility to congenital scoliosis in Chinese Han population. Spine (Phila Pa 1976). 2014;39:1785–91.
43. Windpassinger C, Piard J, Bonnard C, et al. CDK10 mutations in humans and mice cause severe growth retardation, spine malformations, and developmental delays. Am J Hum Genet. 2017;101:391–403.
44. Thomsen B, Horn P, Panitz F, et al. A missense mutation in the bovine SLC35A3 gene, encoding a UDP-N-acetylglucosamine transporter, causes complex vertebral malformation. Genome Res. 2006;16:97–105.
45. Duncan RJ, Carrig CB, Agerholm JS, et al. Complex vertebral malformation in a Holstein calf: report of a case in the USA. J Vet Diagn Investig. 2001;13:333–6.
46. Edmondson AC, Bedoukian EC, Deardorff MA, et al. A human case of SLC35A3-related skeletal dysplasia. Am J Med Genet A. 2017;173:2758–62.
47. Fei Q, Wu Z, Wang H, et al. The association analysis of TBX6 polymorphism with susceptibility to congenital scoliosis in a Chinese Han population. Spine (Phila Pa 1976). 2010;35:983–8.
48. Papapetrou C, Putt W, Fox M, et al. The human TBX6 gene: cloning and assignment to chromosome 16p11.2. Genomics. 1999;55:238–41.

49. Hubaud A, Pourquié O. Signalling dynamics in vertebrate segmentation. Nat Rev Mol Cell Biol. 2014;15:709–21.
50. Aulehla A, Wehrle C, Brand-Saberi B, et al. Wnt3a plays a major role in the segmentation clock controlling somitogenesis. Dev Cell. 2003;4:395–406.
51. White PH, Chapman DL. Dll1 is a downstream target of Tbx6 in the paraxial mesoderm. Genesis. 2005;42:193–202.
52. Chapman DL, Papaioannou VE. Three neural tubes in mouse embryos with mutations in the T-box gene Tbx6. Nature. 1998;391:695–7.
53. Hirata H, Bessho Y, Kokubu H, et al. Instability of Hes7 protein is crucial for the somite segmentation clock. Nat Genet. 2004;36:750–4.
54. Shimojima K, Inoue T, Fujii Y, et al. A familial 593-kb microdeletion of 16p11.2 associated with mental retardation and hemivertebrae. Eur J Med Genet. 2009;52:433–5.
55. Al-Kateb H, Khanna G, Filges I, et al. Scoliosis and vertebral anomalies: additional abnormal phenotypes associated with chromosome 16p11.2 rearrangement. Am J Med Genet A. 2014;164A:1118–26.
56. Takeda K, Kou I, Kawakami N, et al. Compound heterozygosity for null mutations and a common hypomorphic risk haplotype in TBX6 causes congenital scoliosis. Hum Mutat. 2017;38:317–23.
57. Lefebvre M, Duffourd Y, Jouan T, et al. Autosomal recessive variations of TBX6, from congenital scoliosis to spondylocostal dysostosis. Clin Genet. 2017;91:908–12.
58. Offiah A, Alman B, Cornier AS, et al. Pilot assessment of a radiologic classification system for segmentation defects of the vertebrae. Am J Med Genet A. 2010;152A:1357–71. n/a-n/a
59. Bulman MP, Kusumi K, Frayling TM, et al. Mutations in the human delta homologue, DLL3, cause axial skeletal defects in spondylocostal dysostosis. Nat Genet. 2000;24:438–41.
60. Whittock NV, Sparrow DB, Wouters MA, et al. Mutated MESP2 causes spondylocostal dysostosis in humans. Am J Hum Genet. 2004;74:1249–54.
61. Sparrow DB, Chapman G, Wouters MA, et al. Mutation of the LUNATIC FRINGE gene in humans causes spondylocostal dysostosis with a severe vertebral phenotype. Am J Hum Genet. 2006;78:28–37.
62. Sparrow DB, Guillen-Navarro E, Fatkin D, et al. Mutation of Hairy-and-Enhancer-of-Split-7 in humans causes spondylocostal dysostosis. Hum Mol Genet. 2008;17:3761–6.
63. Sparrow DB, McInerney-Leo A, Gucev ZS, et al. Autosomal dominant spondylocostal dysostosis is caused by mutation in TBX6. Hum Mol Genet. 2013;22:1625–31.
64. Clarke RA, Singh S, McKenzie H, et al. Familial Klippel-Feil syndrome and paracentric inversion inv(8)(q22.2q23.3). Am J Hum Genet. 1995;57:1364–70.
65. Mohamed JY, Faqeih E, Alsiddiky A, et al. Mutations in MEOX1, encoding mesenchyme homeobox 1, cause Klippel-Feil anomaly. Am J Hum Genet. 2013;92:157–61.
66. Malfatti E, Bohm J, Lacene E, et al. A premature stop codon in MYO18B is associated with severe nemaline myopathy with cardiomyopathy. J Neuromuscul Dis. 2015;2:219–27.
67. McGaughran JM, Oates A, Donnai D, et al. Mutations in PAX1 may be associated with Klippel-Feil syndrome. Eur J Hum Genet. 2003;11:468–74.
68. Karaca E, Yuregir OO, Bozdogan ST, et al. Rare variants in the notch signaling pathway describe a novel type of autosomal recessive Klippel-Feil syndrome. Am J Med Genet A. 2015;167:2795–9.
69. Vozzi D, Licastro D, Martelossi S, et al. Alagille syndrome: a new missense mutation detected by whole-exome sequencing in a case previously found to be negative by DHPLC and MLPA. Mol Syndromol. 2013;4:207–10.
70. McDaniell R, Warthen DM, Sanchez-Lara PA, et al. NOTCH2 mutations cause Alagille syndrome, a heterogeneous disorder of the NOTCH signaling pathway. Am J Hum Genet. 2006;79:169–73.
71. Wessels MW, Kuchinka B, Heydanus R, et al. Polyalanine expansion in the ZIC3 gene leading to X-linked heterotaxy with VACTERL association: a new polyalanine disorder? J Med Genet. 2010;47:351–5.

72. Hilger AC, Halbritter J, Pennimpede T, et al. Targeted Resequencing of 29 candidate genes and mouse expression studies implicate ZIC3 and FOXF1 in human VATER/VACTERL Association. Hum Mutat. 2015;36:1150–4.
73. Chung B, Shaffer LG, Keating S, et al. From VACTERL-H to heterotaxy: variable expressivity of ZIC3-related disorders. Am J Med Genet A. 2011;155A:1123–8.
74. Saisawat P, Kohl S, Hilger AC, et al. Whole-exome resequencing reveals recessive mutations in TRAP1 in individuals with CAKUT and VACTERL association. Kidney Int. 2014;85:1310–7.
75. Nakamura Y, Kikugawa S, Seki S, et al. PCSK5 mutation in a patient with the VACTERL association. BMC Res Notes. 2015;8:228.
76. Stankiewicz P, Sen P, Bhatt SS, et al. Genomic and genic deletions of the FOX gene cluster on 16q24.1 and inactivating mutations of FOXF1 cause alveolar capillary dysplasia and other malformations. Am J Hum Genet. 2009;84:780–91.
77. Shi H, Enriquez A, Rapadas M, et al. NAD deficiency, congenital malformations, and niacin supplementation. N Engl J Med. 2017;377:544–52.
78. Keegan CE, Mulliken JB, Wu BL, et al. Townes-Brocks syndrome versus expanded spectrum hemifacial microsomia: review of eight patients and further evidence of a "hot spot" for mutation in the SALL1 gene. Genet Med. 2001;3:310–3.
79. Fischer S, Ludecke HJ, Wieczorek D, et al. Histone acetylation dependent allelic expression imbalance of BAPX1 in patients with the oculo-auriculo-vertebral spectrum. Hum Mol Genet. 2006;15:581–7.
80. Lopez E, Berenguer M, Tingaud-Sequeira A, et al. Mutations in MYT1, encoding the myelin transcription factor 1, are a rare cause of OAVS. J Med Genet. 2016;53:752–60.
81. Gucev ZS, Tasic V, Pop-Jordanova N, et al. Autosomal dominant spondylocostal dysostosis in three generations of a Macedonian family: negative mutation analysis of DLL3, MESP2, HES7, and LFNG. Am J Med Genet A. 2010;152A:1378–82.
82. Saker E, Loukas M, Oskouian RJ, et al. The intriguing history of vertebral fusion anomalies: the Klippel-Feil syndrome. Childs Nerv Syst. 2016;32:1599–602.
83. Tracy MR, Dormans JP, Kusumi K. Klippel-Feil syndrome: clinical features and current understanding of etiology. Clin Orthop Relat Res. 2004;424:183–90.
84. Samartzis DD, Herman J, Lubicky JP, et al. Classification of congenitally fused cervical patterns in Klippel-Feil patients: epidemiology and role in the development of cervical spine-related symptoms. Spine (Phila Pa 1976). 2006;31:E798–804.
85. Alazami AM, Kentab AY, Faqeih E, et al. A novel syndrome of Klippel-Feil anomaly, myopathy, and characteristic facies is linked to a null mutation in MYO18B. J Med Genet. 2015;52:400–4.
86. Skuntz S, Mankoo B, Nguyen MT, et al. Lack of the mesodermal homeodomain protein MEOX1 disrupts sclerotome polarity and leads to a remodeling of the cranio-cervical joints of the axial skeleton. Dev Biol. 2009;332:383–95.
87. Rodrigo I, Bovolenta P, Mankoo BS, et al. Meox homeodomain proteins are required for Bapx1 expression in the sclerotome and activate its transcription by direct binding to its promoter. Mol Cell Biol. 2004;24:2757–66.
88. Ajima R, Akazawa H, Kodama M, et al. Deficiency of Myo18B in mice results in embryonic lethality with cardiac myofibrillar aberrations. Genes Cells. 2008;13:987–99.
89. Dietrich S, Schubert FR, Gruss P. Altered Pax gene expression in murine notochord mutants: the notochord is required to initiate and maintain ventral identity in the somite. Mech Dev. 1993;44:189–207.
90. Dietrich S, Schubert FR, Lumsden A. Control of dorsoventral pattern in the chick paraxial mesoderm. Development. 1997;124:3895–908.
91. Dietrich S, Gruss P. Undulated phenotypes suggest a role of Pax-1 for the development of vertebral and extravertebral structures. Dev Biol. 1995;167:529–48.
92. Turnpenny PD, Ellard S. Alagille syndrome: pathogenesis, diagnosis and management. Eur J Hum Genet. 2012;20:251–7.

93. Laufer-Cahana A, Krantz ID, Bason LD, et al. Alagille syndrome inherited from a phenotypically normal mother with a mosaic 20p microdeletion. Am J Med Genet. 2002;112:190–3.
94. Giannakudis J, Ropke A, Kujat A, et al. Parental mosaicism of JAG1 mutations in families with Alagille syndrome. Eur J Hum Genet. 2001;9:209–16.
95. Munoz-Aguilar G, Domingo-Triado I, Maravall-Llagaria M, et al. Previously undescribed family mutation in the JAG1 gene as a cause for Alagille syndrome. J Pediatr Gastroenterol Nutr. 2017;64:e135–6.
96. Saleh M, Kamath BM, Chitayat D. Alagille syndrome: clinical perspectives. Appl Clin Genet. 2016;9:75–82.
97. Tsai EA, Gilbert MA, Grochowski CM, et al. THBS2 is a candidate modifier of liver disease severity in Alagille syndrome. Cell Mol Gastroenterol Hepatol. 2016;2:663–75.
98. Khoury MJ, Cordero JF, Greenberg F, et al. A population study of the VACTERL association: evidence for its etiologic heterogeneity. Pediatrics. 1983;71:815–20.
99. Solomon BD, Pineda-Alvarez DE, Raam MS, et al. Analysis of component findings in 79 patients diagnosed with VACTERL association. Am J Med Genet A. 2010;152A:2236–44.
100. Chen Y, Liu Z, Chen J, et al. The genetic landscape and clinical implications of vertebral anomalies in VACTERL association. J Med Genet. 2016;53(7):431.
101. Dworschak GC, Draaken M, Marcelis C, et al. De novo 13q deletions in two patients with mild anorectal malformations as part of VATER/VACTERL and VATER/VACTERL-like association and analysis of EFNB2 in patients with anorectal malformations. Am J Med Genet A. 2013;161A:3035–41.
102. Peddibhotla S, Khalifa M, Probst FJ, et al. Expanding the genotype-phenotype correlation in subtelomeric 19p13.3 microdeletions using high resolution clinical chromosomal microarray analysis. Am J Med Genet A. 2013;161A:2953–63.
103. Touliatou V, Fryssira H, Mavrou A, et al. Clinical manifestations in 17 Greek patients with Goldenhar syndrome. Genet Couns. 2006;17:359–70.
104. Amalnath SD, Subrahmanyam DK, Dutta TK, et al. Familial oculoauriculovertebral sequence with lymphoma in one sibling. Am J Med Genet A. 2008;146A:3082–5.
105. Slavotinek AM, Vargervik K. Expanded spectrum of oculo-auriculo-vertebral spectrum with imperforate anus in a male patient who is negative for SALL1 mutations. Clin Dysmorphol. 2011;20:11–4.
106. Kosaki R, Fujimaru R, Samejima H, et al. Wide phenotypic variations within a family with SALL1 mutations: isolated external ear abnormalities to Goldenhar syndrome. Am J Med Genet A. 2007;143A:1087–90.
107. Zielinski D, Markus B, Sheikh M, et al. OTX2 duplication is implicated in hemifacial microsomia. PLoS One. 2014;9:e96788.
108. Beleza-Meireles A, Hart R, Clayton-Smith J, et al. Oculo-auriculo-vertebral spectrum: clinical and molecular analysis of 51 patients. Eur J Med Genet. 2015;58:455–65.
109. Stray-Pedersen A, Sorte HS, Samarakoon P, et al. Primary immunodeficiency diseases: genomic approaches delineate heterogeneous Mendelian disorders. J Allergy Clin Immunol. 2017;139:232–45.
110. Posey JE, Harel T, Liu P, et al. Resolution of disease phenotypes resulting from multilocus genomic variation. N Engl J Med. 2017;376:21–31.
111. Zheng HF, Forgetta V, Hsu YH, et al. Whole-genome sequencing identifies EN1 as a determinant of bone density and fracture. Nature. 2015;526:112–7.

Chapter 5
Animal Models of Idiopathic Scoliosis

Zhaoyang Liu and Ryan Scott Gray

Introduction

The term scoliosis is derived from the Greek skolios (σκολιός), meaning to curve or twist; in human, scoliosis is used to describe any atypical curvatures of the spine greater >10° with rotation of one or more vertebral bodies. In most cases, the term "scoliosis" is commonly used to refer to the so-called (adolescent) idiopathic scoliosis (IS); however, the incidence of abnormal spine curvature can manifest in many human diseases; as of June 2017, there are 774 Online Mendelian Inheritance in Man (OMIM) unique entries with some indication of scoliosis. Given the high incidence of scoliosis among human diseases, it is not surprising that many forms of scoliosis (e.g., congenital, kyphoscoliosis, and IS) are modeled in other vertebrate species. Our focus for this chapter is the review and synthesis of how current animal models of scoliosis inform the pathogenesis of normal spine development, homeostasis, and disease, with particular emphasis on models of scoliosis that develop postnatally without overt vertebral dysplasia. There are detractors of the validity of using animal models to study IS, in some cases declaring that a central tenet of the disease is that it is a strictly "bipedal" or "human" condition. Despite these critiques, we hope to highlight examples where animal models have generated fundamental insights into potential biological origins of IS in humans. Regardless of the ultimate clinical relevance of these models for human IS, we suggest that a deeper mechanistic understanding of spine development, homeostasis, and disease using animal models will broaden the understanding of the molecular genetics of spine development and disease in humans.

Z. Liu · R. S. Gray (✉)
Department of Pediatrics, The University of Texas at Austin Dell Medical School,
Austin, TX, USA
e-mail: ryan.gray@austin.utexas.edu

K. Kusumi, S. L. Dunwoodie (eds.), *The Genetics and Development of Scoliosis*, https://doi.org/10.1007/978-3-319-90149-7_5

Normal Spine Function Requires the Integration of Multiple Musculoskeletal Tissues

The structural units of the spine – the vertebral bodies and the intervertebral discs – are derived from segmented condensations of cartilaginous anlage that originate from the embryonic somites flanking the notochord, which ultimately fuse at the midline [1, 2]. The notochord is derived from chordamesodermal cells during gastrulation and can function as a primitive spine in free-swimming aquatic larvae, such as zebrafish and frog, prior to skeletogenesis [3–5]. In addition to this structural role, the notochord also has a critical role in both formation and structure of the spine as signals derived from this tissue are absolutely required for the formation of a segmented vertebral column [6] and direct formation of the nucleus pulposus portion of the intervertebral discs [7].

The maturation and homeostasis of a healthy, functional spine requires the integration of several musculoskeletal tissues including the bone, cartilage and connective tissue, muscle, and the peripheral nervous system. It stands to reason that overt defects in one or more of these musculoskeletal components of the spine could yield scoliosis, for instance, severe vertebral dysplasia is commonly associated with the spine curvatures. However, in the case of IS, there have been few indications of underlying structural defects that would explain pathogenesis of IS. Thus, it is reasonable to speculate that more subtle, subclinical defects of one or more musculoskeletal components of the spine may contribute to pathogenesis of IS. Indeed, magnetic resonance imaging studies suggest differences in signal intensity in the apex of the curvature, and postmortem analysis has shown that changes in the typical expression of anabolic markers and known markers of disc degeneration are found in the intervertebral disc (IVD) of IS patients [8]. Finally, the biology of how the spine and its integrating components develop during adolescence and are maintained in adults remains poorly understood. For instance, how are the annulus fibrosis and nucleus pulposus components of the IVD maintained and how might subtle defects in these tissues contribute to instability of the spine in IS patients. Moreover, it has been observed that the outer annulus fibrosis of the IVD is innervated in humans [9] and rats [10], yet it remains to be determined how these innervations are important for normal spine physiology and their pathology might contribute to disease.

Dysfunction of extrinsic neuroendocrine factors, such as melatonin [11], or metabolic hormones, such as incretins [12, 13], has been reported to be associated with IS in humans and animal models. While there has been limited mechanistic insight attributed to these associations in vivo, it is very likely that these or other systemic interactions of extrinsic and intrinsic musculoskeletal components of the spine are important for the development and homeostasis of the spine. For these reasons, it is wise to contemplate more diverse, potentially synergistic pathoetiologies for IS. For example, minor defects of innervations of the paraspinal muscles and intervertebral disc might be sufficient to generate instability of the spine during period of rapid growth, leading to scoliosis, as was recently shown in mouse [14]. Alternatively, defects in a key neuroendocrine signaling pathway may lead to decreased bone mineral density/osteopenia which could generate weakness in vertebral endplates generating spine instability. Future research into the pathogenesis of IS should begin to address these hypotheses by empirical testing in animal models. The challenge for these models will be to reconcile these mechanistic insights with the known phenotypes of IS in humans and ultimately confirmed in using precious human tissue samples.

Spine Form Follows Dysfunction

The development of the spine begins in utero but undergoes tremendous growth and maturation during adolescence in humans. The same processes of spine development, postnatal refinement, and maturation of the axial skeleton are recapitulated in other vertebrate species including the mouse and zebrafish model systems. There are three broad classifications of scoliosis in humans including congenital, neuromuscular/syndromic, and idiopathic. Congenital scoliosis (CS) is a developmental disorder characterized by overt structural malformation/dysplasia of one or more vertebral units, which can result in focal spine curvatures, present at birth [15]. Defects in somite segmentation or notochord development have been shown to be primary causes of vertebral malformations and CS in animal models. For example, the majority of mutations associated with human CS disrupt Notch signaling components, many of which have been nicely modeled in the mouse [16, 17]. Whereas, the induction of notochord defects via chemical inhibition of lysyl oxidases [18], genetic disruptions of the extracellular matrix components of the notochord sheath [19–21], or by disruptions of lysosomal-dependent vacuolation of the inner most notochord cells [22] lead to vertebral malformations and scoliosis in the zebrafish model. CS can also be modeled by gene-environment interactions in mouse, as was shown by haploinsufficiency of known Notch signaling components in combination with an environmental stressor, hypoxia, in utero [23]. This is particularly important in light of several observations of increased incidence of IS within families of CS patients [24]. Moreover, recent studies in zebrafish suggest that CS and IS may share a common genetic basis [25, 26], where the pathology is altered by changes in gene dosage of *protein tyrosine kinase 7 (ptk7)* [26] or *ladybird homology domain 1b (lbx1b)* [25] during embryonic development. It will be important to determine whether more subtle defects in somite segmentation or disruptions of notochord development during embryonic development can predispose the onset of IS in humans.

Neuromuscular/syndromic scoliosis encompasses spine defects that are thought to be downstream of a general loss of muscle tone via inherent defects of the axial muscles or of their innervations or initiated by severe joint laxity or general weakness of the connective tissues of the axial skeleton. For instance, patients suffering from Duchenne muscular dystrophy [27], neurodegenerative diseases like Rett syndrome [28], or joint hypermobility/connective tissue diseases such as Marfan syndrome [29] and Ehlers-Danlos syndrome [30] are known to display increased incidence of scoliosis without obvious vertebral malformations in humans. A recent report beautifully illustrates that functional ablation of the TrkC neurons, which provide connections between the proprioceptive mechanoreceptors and the spinal cord, can generate a model of IS in mouse [14]. While animal models exist for a wide-range of neurological and connective tissue disorders, very few examples have been reported to display scoliosis as a phenotype [27, 31, 32]. In our opinion, this may simply be explained by under sampling of the spinal architecture in these models.

In contrast to neurological or syndromic scoliosis, clinical manifestations of IS usually occur during adolescence in otherwise healthy individuals. IS can be viewed as a diagnosis of exclusion, wherein IS patients should not display overt vertebral dysplasia, neurological deficits, or other known diagnoses, although many of these measures of diagnoses are not necessarily tested for in all clinical settings. That said, rare variants in connective genes known to contribute for Marfan and Ehlers-Danlos syndromes are reported to be associated with IS in humans [33, 34], suggesting that some AIS patients could represent subclinical examples of these syndromes. Alternately, the pathogenesis of IS may act via somatic loss of heterozygosity or epigenetic changes in the normal pattern of gene expression of other known CS, neuromuscular, or connective tissue disease genes. Ongoing efforts to utilize modern genomics approaches, coupled with experimental testing in animal models, will be critical to test these complex models of pathogenesis.

What Makes a Good Animal Model of Disease?

The molecular genetics and underlying pathology of IS are not fully understood, despite millennia of clinical investigation [35]. For this reason, it is difficult to understand how progress will be made without well-structured experimental approaches including animal models. In recent years, different animal models displaying characteristics of IS have been characterized. These models represent the most ethical and cost-effective way forward to gain deeper mechanistic understanding of both normal development and disease onset and progression of the spine. An explicit animal model should both model the phenotype(s), underlying genetic causes, and natural history of disease and provide an experimental platform for the development of therapeutic interventions of disease. Unfortunately, there are very few examples of animal models that meet this strict set of criteria for any disease [36]. For these reasons, we see a benefit in considering levels of validity for animal models of human disease [37], both for reviewing the current data and as a way of building toward more relevant animal models of IS in the future. We refrain from making strong assertions of whether an animal model has bona fide relevance to human IS; instead, we will discuss several examples which portray morphological similarities to IS as observed in human; undoubtedly caveats exist. We echo the following metrics of validity to characterize animal models of IS:

Face validity An animal model with similar phenotypic indicators of human IS. At this level, a model should exhibit a postnatal-onset scoliosis, in otherwise healthy individuals, and without obvious vertebral malformations. Indeed, there are multiple examples of animal models of IS induced by surgical resection or tethering of vertebral elements and resection of the pineal gland. Moreover, there are a growing number of heritable, genetic models in mouse and zebrafish that display phenotypes observed in IS (Table 5.1). Undoubtedly, many of these models will fall short of relevance for human IS; however, all of these models provide a foundation for assessing structural principles and biological mechanisms for spine stability.

Construct validity An animal model which displays face validity and also displays a similar underlying molecular genetic basis of IS in humans. At this level, the animal model should be the result of an analogous genetic mutation or relevant cellular pathology that is associated with IS in humans. Thus far, only a few examples of animal models at this level of validity have been reported for IS (Table 5.1). Moving forward the method of genetic engineering utilized to engineer the model should be considered. For instance, a complete loss-of-function, "knockout" model of a candidate IS gene might not be as valid in comparison to an animal model that is engineered to contain a tissue-specific conditional loss-of-function or by engineering a "humanized" allele of a candidate IS mutation. With the advent of modern genome editing and conditional genetic approaches, this level of validity should be the standard for modeling human disease, while less robust genetic approaches such as morpholino "knockdown" or transient transgenics should be avoided.

Predictive validity An animal model that has a similar response to clinically validated therapeutics for IS. Thus far, no animal models of IS have been reported for this level of validity. The development of these "gold-standard" models will be critical for the improvement of current interventions or testing of new therapeutics for IS.

Animal Models

There are four distinct classes of animal models reporting some of the hallmarks of IS: (i) pineal resection models, (ii) mechanical models, (iii) environment models, and (iv) genetic models (Table 5.1). The pineal resection models have been recently, extensively reviewed [11]; for this reason, we will not reassess these models here.

Table 5.1 Animal Models of Idiopathic Scoliosis

Model description	Genetic assay	Induction and specificity of genetic assay	Associated phenotype	Heritable IS model established?	Level of model validity	Association of model with human genetics studies of IS?	Cellular/Tissue defect?	Refs
Guppy	Spontaneous *curveback* mutation	Constitutive	Scoliosis	Yes	F	N/A	N.D.	*Scoliosis*. 2010 Jun 7;5:10.
Medaka	Spontaneous *wavy* mutation	Constitutive	Scoliosis	Yes	F	N/A	Dysplasia of the intervertebral disc	*J Toxicol Pathol* 2016; 29: 115–118
Mouse	Conditional *Gpr126*;*COL2CRE* mutant	Specific to osteochondroprogenitor cells, constitutive, beginning at E12.5	Late-onset thoracic scoliosis with co-incident rib cage malformation	Yes	F, C	Yes	Increased cell death and misregulation of gene expression in costal rib cartilages and IVD tissues	*Hum Mol Genet*. 2015 Aug 1;24(15):4365–73.
Mouse	Conditional *Shp2;COL2CRE*^{ERT2} mutant	Induction in osteochondroprogenitos/chondrocytes at 4 weeks	Thoracic and lumbar scoliosis, kyphosis, with rotation	Yes	F	No	Growth plate defects	*Spine* (Phila PA 1976). 2013 Oct 1;38(21):E1307–12.
Mouse	Heterozygous *Chd2*/+ mutant	Constitutive	Lordokyphosis	Yes	F, C	Yes	N.D.	*Am J Med Genet A*. 2008 May 1;146A [9]:1117–27.

Mouse	Double *Gdf5/6* mutant	Constitutive	Scoliosis	Yes	F	No	Connective tissue; joint fusion, inner ear defects	*Developmental Biology* 254 (2003) 116–130
Mouse	Conditional transgene *Col11a2::NPR2* mutation	Cartilage-specific, constitutive expression, beginning at E11.5	Severe scoliosis	Yes	F	No	Growth plate defects	*PLOS One* vol7, issue 8, Aug 2012
Mouse – Mutant	*Fgfr3 −/−*	Constitutive	Late-onset progressive scoliosis, kyphosis, with rotation	Yes	F	No	Overgrowth of long bone, osteopenia, inner ear defects	*Frontiers in endocrinology* Gao et al., 2015
Rabbit – Hereditary lordoscoliotic rabbit	Spontaneous mutation	Constitutive	Scoliosis	Yes	F	N/A	Increased melatonin levels were associated with mutants	*Spine* (Phila PA 1976). 2003 Mar 15;28(6):554–8.
Zebrafish – Mutant	*ptk7*	Zygotic mutant	Scoliosis	Yes	C	Yes	Ependymal cell cilia defects, cerebral spinal fluid flow defects	*Nat Commun*. 2014 Sep 3;5:4777 and *Science*. 2016 Jun 10;352(6291):1341–4.

(continued)

Table 5.1 (continued)

Model description	Genetic assay	Induction and specificity of genetic assay	Associated phenotype	Heritable IS model established?	Level of model validity	Association of model with human genetics studies of IS?	Cellular/Tissue defect?	Refs
Zebrafish – Mutant	*c21orf59*	Temperature sensitive mutation	Scoliosis	Yes	F	No	Ependymal cell cilia defects, cerebral spinal fluid flow defects	*Science*. 2016 Jun 10;352(6291):1341–4.
Zebrafish – Mutants	*ccdc40, ccdc151, dyx1c1*	Constitutive, after RNA injection to rescue embryonic phenotypes	Scoliosis	Yes	F	No	Ependymal cell cilia defects, cerebral spinal fluid flow defects	*Science*. 2016 Jun 10;352(6291):1341–4.
Zebrafish – Mutant	*kif6*	Constitutive	Scoliosis	Yes	F	No	No overt cilia defects in embryos; N.D.	*Dev Dyn*. 2014 Dec;243(12):1646–57.
Zebrafish – Overexpression/ knockdown	*poc5*	Morpholino injection, followed by expression of human wild-type or candidate mutations of POC5	Humanized *POC5* expression was unable to rescue morphants	No	C	Yes	N.D.	*J Clin Invest*. 2015 mar 2;125(3):1124–8.

Zebrafish – Overexpression	*lbx1b*	Overexpression of *lbx* genes. Transgenic overexpression of *lbx1b* from the *lbx1b* promoter	Overexpression did not allow for adult viability. Most Tg(GATA2-1b:lbx1b) founder fish displayed CS-like defects, some displayed IS-like spine curvatures	No	F	Yes	N.D.	*PLOS Genetics* January 28, 2016
Zebrafish – Mosaic mutations; F0 CRISPR	*mapk7*	F0, mosaic CRIPSR mutations	Scoliosis	Not reported	C	Yes	N.D.	*Human Mutation*. 2017;1–11.

F face validity, *C* construct validity

Mechanical Models of IS

The majority of IS models reported thus far have been generated by the establishment of a mechanical asymmetry via bracing, tethering, or resection of axial tissues. It is clear that bipedal and quadrupedal animals do not share identical vectors of axial loading and stress; however, all vertebrates experience mechanical loads and torsion on the spine as they move through the environment regardless of whether they are terrestrial or aquatic animals [38]. By systematically altering mechanical properties of the axial column, these studies have provided critical insights of the anatomical components necessary for normal spine stability. Furthermore, the majority of these experimental approaches have been shown to destabilize quadrupedal animals which further supports the notion that these principles are evolutionarily conserved biomechanical properties of spine stability, irrespective of the forces applied to the spine during locomotion.

The first reports of experimental scoliosis by systematic resection analysis were performed in rabbit by Langenskiold and Michelsson [39–41]. In these seminal experiments, the authors provide an anatomical framework for spine stability, concluding that "... unilateral resection of the posterior ends of the sixth to eleventh ribs including the costal parts of costo-ventral joints" provokes a model of progressive scoliosis. This suggests that the normal attachment of the lower rib cage to the vertebral column is a critical biomechanical component for maintaining spine stability. In agreement, unilateral rib osteotomy or nonsurgical rib cage deformity via thoracic restraint prior skeletal maturity was also shown to generate progressive thoracic scoliosis with rotation in mouse [42]. Importantly, the induction of scoliosis in this thoracic restraint model was relieved by bilateral rib neck osteotomy prior to bracing. Together these data support a model where the biomechanical decoupling of the rib cage and spine can lead to onset and progression of scoliosis, as had been previously proposed [43, 44]. For this reason, abnormal interactions of the spine and rib cage should be considered as a principle etiological factor causing progressive thoracic scoliosis in humans. This model is further supported by observations in that IS patients which commonly have compressed, flatten rib cage morphology or disruptions of sternum [35, 45–49]. Longitudinal studies of human IS to assess how changes in the morphology of the rib cage contributes to the onset and progression of IS will be greatly assisted by modern low-dose radiation, 3D imaging platforms [50].

Postnatal spine maturation proceeds via the growth and ossification of several cartilaginous unions (synchondrosis) to include the costovertebral joints – which articulate each rib with an individual vertebral unit – and the neurocentral joints, which join the neural arch to the vertebral body or centrum. Moreover, the vertebral units, which are derived from cartilaginous anlage flanking the notochord, are themselves synchondroses, typically fused prior to birth in humans. Perhaps defects in the maturation and closure of these axial midline synchondroses contribute to the onset of IS. As described above the ablation of the normal mechanical properties of the costovertebral joints can generate robust IS-like scoliosis in rabbit and mouse.

In agreement the hemi-circumferential ablation of the neurocentral joint by electrocoagulation results in mild scoliosis in a growing pig model [51]. Taken together it is clear that the abnormal development and maturation of these axial synchondroses may serve as a source of mechanical asymmetry, contributing to the onset of IS in some cases.

In addition to intrinsic defects of the spine, extrinsic components of the spine such as dysfunctional sensorimotor control, asymmetry of the semicircular canal, and vestibular defects have been proposed to contribute to pathogenesis of IS [52–56]. Resection studies in the frog *Xenopus laevis* show that unilateral resection of the inner ear/vestibular structures results in progressive scoliosis without vertebral dysplasia [57]. The spine curvatures in this model are hypothesized to be due to the introduction of an asymmetric muscle tone and a progressive deformation of the cartilaginous structural elements of the spine prior to ossification [58]. The model born out of these findings suggests that loss of vestibular function may promote a general tonic imbalance in descending pathways of spinal motoroneurons which would contribute to asymmetric torsion on the spine; in particular during development in utero, where weight-supporting limb proprioceptive signals are diminished and the immature skeleton is composed of less ossified structural elements, connective tissues, and synchondrosis. In this manner, even subtle defects of the biomechanical properties of these elements of the spine could be primary to the initiation of scoliosis during periods of rapid growth in adolescence. On the other hand, it may be possible that any imbalance of descending neurological pathways during embryonic and postnatal development, whether peripheral or central, may initiate deformations of the soft spine elements via asymmetric muscle tone. Many disorders affecting the central nervous system in humans also manifest scoliosis; however, several mouse models with defects in the peripheral nervous system do not display scoliosis [59–62]. However, it was recently shown that tissue-specific ablation of *Runx3* from peripheral neurons (*Wnt-1Cre*), which ablates proprioceptive function, generates characteristics of IS with generalized ataxia in mouse [14]. In conclusion, these mechanical animal models of scoliosis provide strong support that both structural and neuromuscular asymmetries during periods of spine development can cause scoliosis that model IS with face validity.

Environmental Models

There is a report of scoliosis in a wild fish population in Belews Lake, North Carolina. These fish displayed curves along both dorsal-ventral and medial-lateral axes, which is phenotypically similar to scoliosis reported in other zebrafish mutants with late-onset scoliosis [63, 64]. The cause of scoliosis in these fish is thought to be due to contaminated wastewater effluent from a coal-fired power plant, found to contain high levels of selenium. Interestingly, after bioremediation efforts which decreased levels of selenium, the incidence of spine deformity was found to be concomitantly decreased in the population [65]. This report suggests that high levels of selenium have teratogenic effects on normal spine development. The mechanism of this effect has not been determined or recapitulated in other established laboratory animal models.

Genetic/Heritable Models of IS

There are multiple examples of both aquatic and terrestrial animals in the wild which display IS-like scoliosis, including bonobo [66], orangutan [67], dolphins [68, 69], gray whale [70], and sea otter [71], although it is unknown whether the pathogenesis of these curvatures represents environment or genetic factors. Regardless, there are several examples of heritable models that display characteristics of IS reported in guppy, quail, chicken, and rabbit. Unfortunately, most of these animal models are no longer available for modern genetic studies [72–74]. More recently studies using genetically defined laboratory animal models have begun to identify heritable genetic lesions which generate models of IS [26, 62, 63, 75]. The continuation of mechanistic studies in these genetically defined, heritable animal models will greatly aid in the deeper understanding of the genetics of normal spine development and will contribute to our understanding of the pathogenesis of scoliosis in humans.

Aquatic Models of IS

The spontaneous *curveback* mutant guppy (*Poecilia reticulata*) model of IS which displays face validity due to larval-onset scoliosis with a dorsal-ventral sinusoidal curvature, with additional medial-lateral curves and rotation, without the presence of any underlying vertebral malformations [76]. This mutant phenotype was mapped to a qualitative trait locus on the guppy linkage group 14, which was estimated to explain >80% of the genetic variance [77]. The exact nature of the genetic lesion controlling spine instability in the *curveback* mutant has not been reported.

Several other IS models have been reported in the zebrafish (*Danio rerio*) model. The zebrafish model system is one of the premier model organisms for unbiased forward genetic screening of relevant phenotypes as well as for the application of functional testing of genetic associations of human disease using modern gene-editing techniques. Robust techniques have been successfully employed in the zebrafish model to study development and human disease including injection of synthetic RNA constructs, chemical genetic screens, modern transgenic approaches, and the use of antisense morpholino oligonucleotides (morpholinos) to reduced or knockdown gene expression or translation of gene products [78]. With the recent advent of modern reverse genetic approaches such as TALENs or CRISPR-Cas9 [79, 80], it is now feasible to substitute wild-type (WT) alleles with variant, "humanized" alleles found in human genetic studies, making this model a cost-effective resource for the development of animal model of IS displaying construct validity. Moreover, zebrafish undergo external development and have mostly transparent bodies during the range of skeletal development making them increasingly valuable for studying postembryonic skeletal development and disease [21, 64, 81–86].

For these reasons, recent genome-wide association studies (GWAS) of human IS cohorts have utilized the zebrafish model to test for functional roles of these risk loci. For example, the human variant, rs6570507, located within an intron of the *G protein-coupled receptor 126* (*GPR126*) locus, is associated with human IS [87]. To test a model of loss-of-function for *gpr126* in zebrafish, morpholinos were used to knockdown *gpr126* expression in embryonic zebrafish. This resulted in minor defects of the ossification of the axial skeleton in early larval stages [87]; however, no scoliosis was observed in adult mineralized spines in this fish. At first glance this *gpr126* morpholino-based model appears to have some mechanistic validity in the spine; however, it is important to note that several missense and nonsense alleles of *gpr126* in zebrafish have been described to exhibit defects in myelination due to defects in Schwann cell biology [88] and dysplasia of the cartilaginous semicircular canal of the inner ear [89]; in contrast none of these *gpr126* mutants were reported to display defects in bone mineralization or scoliosis. Perhaps the simplest interpretation is that *gpr126* knockdown reagents are exhibiting non-specific effects on the mineralization of the axial centra, having little to do with the true nature of *gpr126* function in zebrafish development.

The human variant, rs11190870, represents one of the most robust signals for risk of IS in multiple ethnic backgrounds [90–93]. This SNP is located several kilo bases downstream of the *Ladybird Homology Domain 1* (*LBX1*) gene. Lbx proteins are homeodomain DNA-binding nucleoproteins with well-defined roles in transcriptional regulation and cell lineage determination including migration of muscle lineages [94] and specification of neuronal subtypes in the brain and spinal cord [95–97]. Guo et al. found increased binding of nucleoproteins as well as increased transcriptional activity using a region of DNA containing the "risk" rs11190870 T-allele as compared to the same region of DNA with the "non-risk" C-allele [25]. Together these findings suggest a gain-of-function model for the T-allele, perhaps leading to increased or ectopic expression of the *LBX1* gene. However, it should be pointed out that both T- and C-alleles are common in a diverse range of ethnic backgrounds. Moreover, the T/T and T/C haplotypes are the most common in all ethnic backgrounds tested which implicates the T-allele as the ancestral allele. In agreement, we have only observed the T-allele in the mouse and zebrafish models, where we have assayed thus far (unpublished data).

The zebrafish model was utilized to test this gain-of-function model for *LBX1* in the pathogenesis of IS. Zebrafish have three paralogues of *LBX1*, *lbx1a*, *lbx1b*, and *lbx2*, and a *LBX1 antisense RNA 1* (*LBX1-AS1*; *FLJ41350*) gene closely linked to the *lbx1a* locus. Scoliosis or axis defects were not observed in either *lbx1b*$^{-/-}$ or *lbx2*$^{-/-}$ mutants; however, neither *lbx1a* mutants nor compound mutants were assayed. To test for a gain-of-function of these genes, the authors injected synthetic RNAs of each of the three zebrafish *LBX1*-related genes, human *LBX1*, or *LBX1-AS1* RNA finding that overexpression of all zebrafish and human paralogues induced embryonic defects including notochord bending and convergent and extension defects of the axial mesoderm. The overexpression of *Lbx* RNAs induced decreased expression of *wnt5b* in the presomitic mesoderm, which is necessary for convergent and extension behaviors during zebrafish gastrulation [98]; however, experiments did not yield viable adult fish for analysis of the spine.

Interestingly, CS-like defects and kinks and bends of the notochord and vertebral malformations and scoliosis in mineralized adult spines were observed in transient transgenic zebrafish engineered to express *lbx1b* under the control of a portion *lbx1b* regulatory region (*Tg[lbx1b:lbx1b]*), which faithfully recapitulates the endogenous embryonic expression pattern of *lbx1b*. As previously discussed, structural defects of the notochord are well-established to contribute to vertebral malformations in zebrafish [20–22, 99, 100], which is very likely to be the antecedent of the vertebral malformations in these F0 *Tg[lbx1b:lbx1b]* transgenic zebrafish. Interestingly, a few F0 *Tg[lbx1b:lbx1b]* transgenic animals displayed scoliosis without vertebral malformations, more reminiscent of IS, presumably due to differences in the efficiency of integration or size of the clones expressing the transgene. Alternately, the locus of integration may regulate the spatial and temporal expression of the transgene in turn regulating the onset and severity of the phenotype.

In general, somatic transgenes (F0 s) are not generally considered a valid method for genetic analysis in the zebrafish community, and thus any conclusions made via this type of transient analysis are difficult to interpret. For instance, these defects could be related to non-specific effects due to the overexpression of *lbx1b*, rather than revealing a relevant function of *lbx* genes during zebrafish spine development. Unfortunately, this approach produced no viable F1 transgenic lines precluding analysis of the mature spine in stable *Tg[lbx1b:lbx1b]* transgenics. These data illustrate that expression of *lbx1b* must be tightly regulated during normal development in zebrafish and that dysregulation of *LBX1* expression may highlight a common genetic basis of CS and IS in humans.

Additional support for a genetic basis for CS and IS was described by loss-of-function studies of the *ptk7* gene in zebrafish [26], which showed that maternal-zygotic *ptk7* mutants (*MZptk7*) – where the eggs are lacking *ptk7* gene product – display vertebral anomalies commonly associated with CS (e.g., hemivertebrae and vertebral fusions). Mechanistically, *MZptk7* mutant embryos display defects in the patterning of known somite segmentation pathway genes and defective Wnt signaling, which is well known to be important during somitogenesis [101]. In contrast, zygotic *ptk7* (*Zptk7*) mutant zebrafish – where the egg contains maternally deposited wild-type *ptk7* gene products but has little or no functional zygotic transcription of *ptk7* – displayed defects more characteristic of IS. Importantly, a single rare variant of *PTK7* (*P545A*) isolated in a patient with IS was shown to generate defects in Wnt signaling. The pathogenicity of this PTK7 coding variant has not been directly tested in an animal model. *Zptk7* mutant zebrafish display strong face validity with several hallmarks of IS observed in humans including postnatal-onset scoliosis and lack of vertebral dysplasia and in animals that display no apparent defects in behavior or viability.

Further analysis demonstrated that *ptk7* expression in *foxj1a*-postive cells or tissues (Tg[*foxj1a::ptk7*]) is sufficient for maintaining spine stability during larval development [75]. *Foxj1* is a master transcriptional regulator of motile cilia [102], and the motile ependymal cell cilia lining the ventricles are thought to help in the establishment of normal function of the ventricular system by helping to facilitate flow of the cerebral spinal fluid (CSF) in the brain. Importantly, a stable *Tg[foxj1a::ptk7]* transgene rescued the loss of ependymal cell cilia and normal CSF flow in *Zptk7* mutant zebrafish. Together these data implicate defects in ependymal cell cilia and CSF flow in the pathogenesis of IS. Similar defects are typically associated with hydrocephalus and postnatal lethality in mouse; it will be interesting to see if milder defects in the ventricular system would affect spine stability. It will be important to determine if defects in the ventricular system of the brain and spinal cord might underlie the association of scoliosis that is secondary to many neurological disorders such as Rett syndrome and cerebral palsy.

The theme of ependymal cell cilia and ventricular defects contributing to scoliosis in zebrafish is recapitulated by several other mutants which disrupt components of primary or motile cilia, CSF flow, or generate primary ciliary dyskinesia in zebrafish [75]. In particular, a temperature-sensitive allele (*tm304*) of the *c21orf59* gene was used to define a temporal window of larval development (from 18 to 30 days' postfertilization (dpf)) when its function is required to maintain normal spine morphology. Presumably, these experiments also define a critical window when the function of these motile cilia components and functional CSF flow is necessary for spine homeostasis; however, the precise link between the ventricular system and spine stability remains to be determined. Interestingly, this period of development in zebrafish corresponds to well-documented periods of rapid growth in zebrafish [26, 103] and may be analogous to periods of rapid growth during adolescence in humans. These mutant zebrafish models of IS exhibit face validity and provide novel models of defective CSF flow in zebrafish with which to investigate how this process may regulate typical spine morphology.

Finally, the relationship between inadequate CSF flow and scoliosis has been observed in other experimental models of syringomyelia, generated by injection of Kaolin (hydrated aluminum silicate) into the subarachnoid space in rabbit and dog [104, 105] and may underlie spine defects humans with Chiari malformations which also display obstructed CSF flow and higher rates of scoliosis [106]. It will be interesting to understand if subclinical deficits in ependymal cell cilia or alterations to CSF flow may contribute to risk of IS in humans.

Other genes predicted to encode proteins that are core components of the cilia have been implicated in IS in zebrafish including the *kinesin family member 6* (*kif6*) gene, a microtubule motor protein, and the centriolar protein homolog *poc5*. In the case of *kif6,* multiple non-complementing frameshift mutations generate late-onset scoliosis without vertebral dysplasia [63], which provides a model of face validity for IS. While no defects in ciliogenesis of either primary or motile cilia were observed in *kif6* mutant embryos up to 5dpf, the *kif6* transcript was shown to be expressed in the brain. It will be interesting to test whether *kif6* has a role in the ventricular system as was shown for *ptk7*. While there are no associations reported as yet between *KIF6* and IS in humans have been reported thus far, a recent report suggests that a micro RNA, *MIR4300HG*, associated with progression of IS in human may bind *kif6* transcript, presumably to control its regulation [107].

Several familial cohorts containing IS are reported to be associated with rare single nucleotide variants (SNVs) of the *POC5* gene [108]. Using morpholino antisense oligonucleotides (MO), the authors observed that knockdown of zebrafish *poc5* generated a curly-tailed phenotype with lethality after 3dpf phenotypes commonly associated with primary cilia dyskinesia in zebrafish [109–112]. A *poc5* MO was used to "knockdown" endogenous *poc5* expression, and human RNAs were co-injected to examine the effects of candidate human *POC5* SNVs during spine development. Embryos expressing mutant *POC5* RNAs were viable but were observed to have mild to severe axial curvatures in larval fish and IS in the fully mineralized adult skeleton, whereas embryos injected with WT human *POC5* RNA did not develop scoliosis, suggesting that the IS-associated SNVs of *POC5* are loss-of-function alleles [108]. The frequency of these SNVs of *POC5* in these human cohorts was 75% and 30% in each of the cohorts, suggesting that these variants of *POC5* are not sufficient to generate pathology in humans, rather there are likely to be additional disease-modifying mutations that co-segregate with effected individuals. It remains to be determined if the generation of endogenous "humanized" alleles of *poc5* would display similar defects as were observed by knockdown and overexpression of exogenous *POC5* variants. The generation of these alleles may provide construct validity and would be a valuable resource for modifier screens to identify IS loci in the zebrafish model.

Zebrafish genetics provide powerful models for validation for human genetic studies. Many important advances in developmental biology have been made by careful use of morpholino oligonucleotide-based gene interference [113]; however, there are many examples where lack of sufficient controls using MOs confounds the interpretation of the results. Along the same lines, overexpression by RNA injection or by transgenesis in zebrafish allows robust functional assays of human genetic findings. However, careful analysis using dose curves for injection of RNA or generation of stable transgenic lines should be the standard for all experimental design in future studies.

Mouse Models

There are several mutant mouse strains that model aspects of IS, with the majority of these being the result of defects in the development or homeostasis of connective tissues and cartilages (Table 5.1). Using the inducible Col2a1CreERt2 deleter strain, postnatal conditional loss of the *protein tyrosine phosphatase, non-receptor-type 11* gene (also known as *Shp2*) in chondrocytes (*Col2a1+* lineages) during juvenile development (4 weeks of age) generates a late-onset scoliosis in about 40% of the *SHP2*-conditional mutant mice by the age of 12 weeks [114]. The spine curvature in this conditional mutant mouse was very severe, noticeable without X-ray, atypical lordosis in the upper thoracic spine and kyphosis of the lower thoracic to lumbar spine as well as rotation of individual vertebrae. SHP2 is a positive regulator of RAS-MAPK signaling, which is essential for connective tissue growth [115], and this conditional mouse model of *Shp2* suggests this signaling is required for homeostasis of the intervertebral disc. Indeed, histological analysis revealed variable thickness and a disruption of typically aligned columnar chondrocytes in the vertebral growth plate in these conditional mutant mice. Interestingly, when *Shp2* was removed from Col2a1-expressing lineages later in development (8 weeks of age), no spinal deformity or scoliosis was observed up to 16 weeks of age. The analysis of this conditional *Shp2* mutant mouse reveal a prospective window of susceptibility for pathogenesis of scoliosis due to defects of the disc and that homeostasis of cartilage tissues can be a major driver of spine stability.

Additional associations between cartilage homeostasis and spine integrity are underscored cartilage-specific deletion of SOX9 in postnatal mice, using an inducible aggrecan enhancer-driven CRE deleter strain (*Agc1-CreERT2*) [116]. SOX9 is an essential transcription factor for the differentiation of the chondrocyte lineages during embryonic development [117]. In order to circumvent the embryonic requirement for SOX9, recombination of a floxed *Sox9* allele was induced at 6 weeks of age, and by 4 months the mice displayed severe kyphosis of the thoracic spine. This postnatal loss of *Sox9* also resulted in compression and degeneration of the IVDs, depletion of sulfated proteoglycan and aggrecan content in IVD cartilage, and precocious growth plate closure, and the mutant mice were observed to have defects in overall growth, although the size of individual vertebrae in adult mice was not affected. Global RNA transcriptome sequencing (RNA-seq) analysis revealed that depletion of SOX9 in aggrecan-expressing lineages of the IVD remarkably reduced the expression of several genes encoding extracellular matrix proteins, as well as some enzymes responsible for their posttranslational modification. Furthermore, several cytokines, cell-surface receptors, and ion channels were dysregulated, confirming that SOX9 also has a critical role in coordinating the homeostasis of the IVD postnatally and which is likely to contribute to loss of spine stability. Similar testing at 8 weeks was not reported in this model so it is unclear if spine stability is resistant to a loss of *Sox9* expression at later time points as was observed for *Shp2*.

The regulation of spine structure during postnatal development is also controlled by the growth differentiation factors 5 and 6 (*Gdf5* and *Gdf6*) genes, wherein *Gdf5/6* double mutant mice display severe vertebral column defects in adult mice that are not seen in either *Gdf5* or *Gdf6* single mutant animals [118]. While most double mutant mice fail to thrive to adulthood, severe lateral curvatures (Cobb angle between 39 and 67 degrees) of the spine were observed in roughly 30% of the double knockout mice by the age of 3 months, without vertebral dysplasia. *Gdf5* is expressed in the developing joints of the skeleton and is one of the earliest known markers of joint formation [118, 119]. Histological analysis of the spine and IVD in these double mutant mice reveals reduced proteoglycan staining as indicating a degeneration phenotype of the cartilage. More global defects in the skeletal system were also observed, including a reduction in mineralized bone and narrowing or loss joint space in the limbs, which are not common traits of IS in humans. Despite this these data illustrate that both *Gdf5* and *Gdf6* are candidate risk loci for the pathogenesis and progression of IS.

While these models all display some face validity for understanding IS, the overt nature of other co-occurring pathologies makes these models less than ideal. Regardless it is important to underscore that these mutant mouse models are likely due to strong loss-of-function models, which may not be the case in human IS. Alternatively, it is more likely that hypomorphic mutations, compound heterozygous mutations, or extragenic enhancer mutations that reduce the level or pattern of expression of these genes will provide some better models of IS. For this reason, variants in the genes highlighted above as well as other loci of well-known genes important for cartilage biology, especially those containing genes known to be important for cartilage, connective tissue, or IVD development, should be considered for validation in animal models.

Changes in bone mineral density or osteopenia may contribute to risk of IS in humans [120, 121]. Interestingly, loss-of-function mutant mice of the *Fibroblast growth factor receptor 3* (*Fgfr3*) gene display scoliosis and kyphosis in juvenile *Fgfr3*$^{-/-}$ mice, which is associated with overgrowth of the axial and appendicular skeleton [122, 123]. *Fgfr3*$^{-/-}$ mice have also been characterized to display reduced cortical bone thickness, defective trabecular bone mineralization, premature joint degeneration, and early arthritis [124, 125]. Longitudinal X-ray and Micro-CT analysis of individual *Fgfr3* mutant mice shows that this model displays progressive onset of spine curvature, where rapid progression in the thoracic spine coincided with rapid body growth up to 4 months [126]. Additional analysis of these mutant spines revealed lateral displacement of the spine and axial rotation of the vertebrae in *Fgfr3* mutants. In addition, an overall increase in the length of vertebrae and a significant reduction of height on the concave side of IVD or wedging were also observed. Quantitative analysis of vertebrae, outside of the curve, showed reduced trabecular bone volume. Interestingly, higher measures of bone mineral density were observed at the concave side compared with the convex side vertebrae within the curve. This phenotype may result from an anabolic response to increased biomechanical strain on the concave side [127]; alternatively, this could be the result of a collapse of the disorganized trabecular network under increased strain [128].

Local treatment of the thoracic spine with a bone anabolic agent PTHrP-1-34 inhibited the progression of lateral scoliosis and vertebral wedging phenotypes in *Fgfr3* mutant mice, but this treatment had little impact on the formation of kyphosis in this model. The changes in IVD morphology in these *Fgfr3* mutant mice were also corrected with PTHrP-1-34 treatment. Taken together, these results suggest that FGFR3 plays a critical role in regulating bone and cartilage growth, and PTHrP activity might be complementary to a loss of FGF signaling in the spine. While this *Fgfr3* mutant model displays strong face validity for IS, no associations with this locus have been observed in humans with IS. That said, the mutations uncovered in the human *FGFR3* locus have been dominant mutations and are known to cause a variety of limb malformations or cranial bone dysplasia – with one exception, where a novel pathogenic homozygous missense mutation in *FGFR3* was isolated in a cohort of patients characterized by tall stature and scoliosis. While this is an interesting validation of the potential role of *FGFR3* for pathogenesis of IS, other abnormalities of the appendicular and axial skeletons in these patients preclude this from being clinically classified as IS; instead this condition is labeled as a skeletal overgrowth syndrome [129]. Regardless, these studies illustrate that FGFR3 function is required for postnatal regulation of spine development and stability and further investigation of this locus and of FGFR3 signaling in human IS patients is warranted.

The C-type natriuretic protein (CNP) and its receptor, *natriuretic peptide receptor 2* (*Npr2*), and downstream effectors are involved in long bone growth in mice [130, 131]. Several gain-of-function mutations of *NPR2* have been identified in patients with an overgrowth syndrome characterized by tall stature, macrodactyly, and scoliosis [132, 133]. Interestingly, transgenic mice strain that expresses a gain-of-function allele of *NPR2* (p.Val883Met) from a cartilage-specific *Col1a1* promoter was observed to display kyphoscoliosis, wider growth plates, increased bone length, as well as upregulation of cyclic guanosine monophosphate (cGMP) level in cartilage [133]. While these alleles in human generate a broader spectrum of phenotypes than commonly observed in IS in human, the mechanism of cartilage overgrowth potentially via over production of cGMP and its downstream effectors may be a relevant mechanism of IS to consider. This mouse model highlights a successful approach in using transgenic mouse genetics to test gain-of-function models of human disease, which may be applied to study IS in the future.

Perhaps the most relevant model of IS has been observed as a conditional loss-of-function of *Gpr126* specifically deleting in osteochondroprogenitor cells (using the Collagen Type II Cre (Col2Cre) deleter strain) [62]. It was observed that embryonic loss of *Gpr126* in osteochondroprogenitor cells resulted in postnatal-onset of spine curvature without vertebral dysplasia in ~50% of the *Gpr126* conditional knockout mice by the age of postnatal day 20 and with increased incidence (>85%) by 4 months of age. Because this *Col2Cre;Gpr126* mutant mouse model has many hallmarks of IS including lack of other vertebral dysplasia and postnatal-onset of the pathology (Fig. 5.1), and because the *GPR126* locus is associated with human IS [87], we suggest it should be viewed as a model of IS with construct validity.

GPR126 has been shown to be required in multiple tissues including during endocardium development [134], in Schwan cells for the myelination of peripheral axons [88, 135], and for inner ear development [89]. The exact cellular etiology of the *Col2Cre;Gpr126* mutant mouse has not been established; however, it is clear that osteochondroprogenitor cells give rise to the bone, cartilages and connective tissues, the IVD, and to many of the tendons and ligaments of the spine. Further work using more refined conditional mouse genetics is needed to address which of these tissues is critical for the pathogenesis of IS in this model.

That said histological analysis of the IVDs did not reveal any overt changes in the patterning or differentiation of the IVD tissues a P1 or P20 in *Col2Cre;Gpr126* mutant mice, with the exception of some incidence of acellular clefts at the midline of the annulus fibrosis and growth plate at both P1 and P20. This suggests that *Gpr126* may have a role in the normal closure of the midline axial synchondroses of the vertebral bodies during the transition from notochord to IVD. Moreover, mild

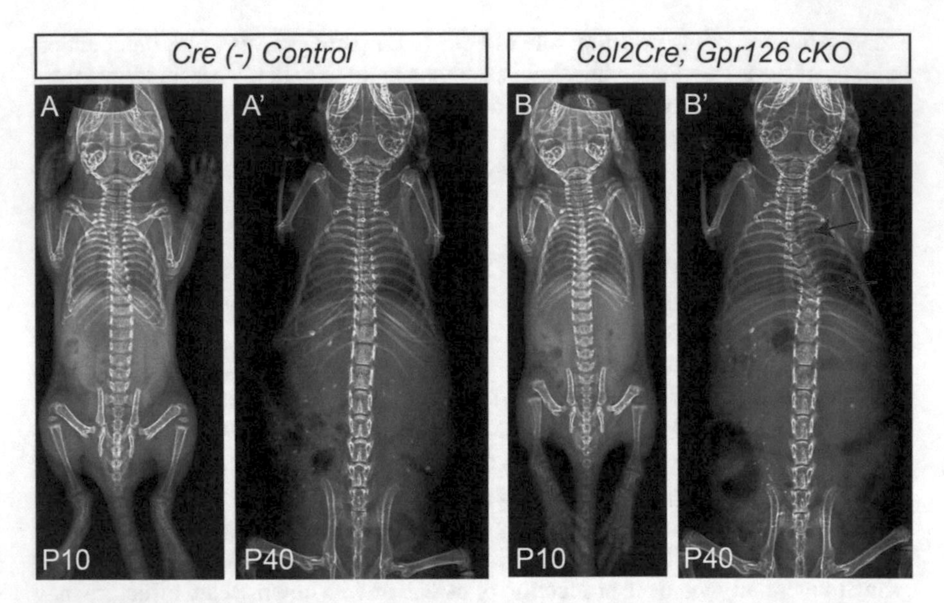

Fig. 5.1 Col2Cre;Gpr126 mutant mice display late-onset scoliosis without vertebral dysplasia. Longitudinal X-ray imaging of (dorsal view) of an individual Cre(-) control (**A-A'**) and *Col2Cre;Gpr126* mutant (**B-B'**) mutant mice. The mice of both genotypes are phenotypically normal without overt vertebral dysplasia at postnatal day 10 (P10); however, by P40 the *Col2Cre;Gpr126* mice display right thoracic spine curvature (red arrows) that are not observed in Cre (-) control mice

increases in cell death visualized by increased TUNEL-positive cells of the vertebral growth plate and IVD were also observed, suggesting a minor role of *Gpr126* in cell survival in these tissues. In contrast, no significant changes in the trabecular bone of the vertebrae or defects in the development or mineralization of the long bones was observed in *Col2Cre;Gpr126* mutant mice, suggesting that *Gpr126* functions in chondrocytes or other connective tissues and not in the bone to maintain the homeostasis of the spine. However, the use of well-characterized lineage-specific CRE transgenic mouse strains will be necessary to support this model. Additionally, further studies should also seek to address if the scoliosis observed in conditional mouse mutants of *Gpr126* is the result of embryonic defects or if its function is required during periods of rapid spine growth.

Interestingly, a significant fraction of these *Gpr126* conditional mutant mice also exhibited dorsal-ward deflections of the sternum, reminiscent of rib cage defects clinically termed *pectus excavatum* (PE) in humans. PE is a common musculoskeletal disorder of the anterior chest wall which has high concomitant incidence with IS in humans [46, 47, 136]. The authors found that loss of *Gpr126* led to upregulated expression of a matrix modifying gene *Galactose-3-O-sulfatransferase* (*Gal3st4*), a gene implicated in human PE [137]. Taken together, these data suggest that *Gpr126* may act as a common genetic cause for the pathogenesis of both IS and

PE, possibly via a misregulation of extracellular matrix gene expression, which was also observed in the inner ear cartilages of *gpr126* mutant zebrafish [89].

Distinct Elements of the Mammalian Spine

While both mouse and zebrafish models are being utilized to study scoliosis, it is important to note distinct differences in the components of the vertebral column between them. For example, the formation of the vertebral units during development and the morphology and composition of the bony vertebrae and of the IVD are quite different. In mouse, the vertebrae form by endochondral ossification of cartilaginous anlagen which fuse at the midline [138]. In contrast, zebrafish vertebrae form by direct mineralization of the notochord sheath to form the perichordal centra which underlie the elaboration of the vertebrate, which does not pass through a cartilaginous stage [139, 140]. Mature mouse vertebrae are columnar-shaped structures that form by endochondral ossification which leaves cavities the interior of the vertebrae, housing the bone marrow. In contrast, mature vertebrae in zebrafish are hourglass-shaped structures, without bone marrow; instead they are filled with vacuolated tissue which is likely derived by the fusion of notochord-derived vacuolated cells [141].

There are also overt differences in the tissue components of the IVD. In mouse, the IVD is a lamellar fibrocartilaginous joint surrounding the nucleus pulposus (NP), characterized by an abundance of hydroscopic proteins (e.g., aggrecan) which act to generate high osmotic pressure, allowing for resilience during compressive strain of the spine [142]. The inner annulus fibrosis (AF) layers of the IVD are composed of strands of fibrocartilage attached to the cartilaginous endplate (CEP). These strands of fibrocartilage in composite form the inner and outer AF lamellar layers that radially and circumferentially surround the NP tissue (Fig. 5.2A, A'), providing structural integrity and containment for the NP [6].

The IVD of zebrafish contains notochord-derived [141] vacuolated cells embedded in a fibrocartilaginous matrix (Fig. 5.2B, B') which highlights a common lineage of the innermost portion of the IVD in both mouse and zebrafish. Despite the common origin of the inner disc tissue (NP), zebrafish do not display some of the molecular hallmarks of NP or AF tissues observed in mouse and human. For instance, Safranin-O staining – which stains mature, healthy AF and NP of sectioned IVD tissue in mouse – does not stain the IVD of adult zebrafish (Fig. 5.2B, B'), in contrast to the cartilage-rich NP of the mouse (Fig. 5.2A, A'). Indeed the formation of glycosaminoglycan-rich NP tissue appears to be a hallmark of mammals [143]. The analogous structure to the AF in zebrafish appears to be a small acellular intervertebral ligament (IVL) that encircles the IVD, originally described in medaka fish (*Oryzias latipes*) [144], and in contrast to the robust fibrocartilaginous strands of cells which make up the AF in mouse (Fig. 5.2A, A'). Interestingly, Safranin-O staining of zebrafish IVD is seen just interior, adjacent to the IVL (Fig. 5.2B', D); however, the function of this group of cells remains unclear.

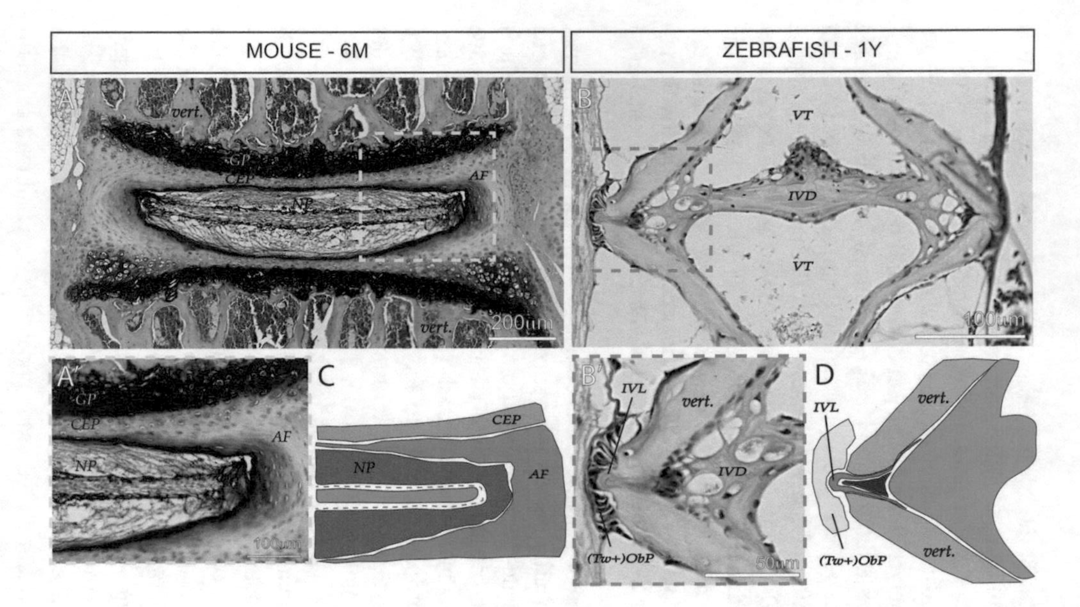

Fig. 5.2 Key differences of the intervertebral disc (IVD) between mouse and zebrafish. Safranin-O/fast green staining of a midline section of a mature intervertebral region in mouse (**A-A'**) and zebrafish (**B-B'**) shows high levels of proteoglycan-rich cartilage in mouse IVD, compared to very little zebrafish. Insets of mouse (**A'**) and zebrafish (**B'**) IVD highlighting the tissue components of the disc. Cartoon schematic of insets for mouse (**C**) and zebrafish (**D**). The mouse IVD displays a large AF which connects two flanking vertebrae at the level of the CEP. The AF in mouse surrounds the NP structure. The NP appears to be composed of three distinct layers: an outer tissue layer which stains for Safranin-O (orange in (**C**)), inner cell layer (dotted red line in (**C**)), and an inner tissue layer that does not stain well for Safranin-O (blue in (**C**)). In contrast, the zebrafish IVD has only weak Safranin-O staining (red in (**B'**, **D**)) in an interior region adjacent to the intervertebral ligament (IVL) (**B'**, green in (**D**)). The zebrafish does not display a true cartilaginous NP tissue, rather the IVD is composed of vacuolated cells and fibrocartilaginous matrix. In contrast to mouse vertebrae which contains bone marrow and trabecular bone filling the vertebrae, zebrafish vertebrae contain bone-shaped vacuolated tissue (VT). The presence of twist-positive osteoblast progenitor cells ((*Tw*+)ObP) are observed adjacent to the IVL. GP growth plate, CEP cartilaginous endplate, NP nucleus pulposus, AF annulus fibrosis, IVL intervertebral ligament, (Tw+)ObP twist-positive osteoblast progenitors, vert. Vertebrae

Zebrafish IVDs are also observed to host a collection of cells adjacent to the IVL on the exterior side, presumably these are *twist*-positive as shown in the medaka fish, hypothesized to be osteoblast progenitors [144], and may also be important for maintaining the IVL tissue. Regardless of these unambiguous differences in the structure and protein composition of mouse and zebrafish IVD, genetic studies that illustrate defects in these tissues, which are correlated to scoliosis, may have conserved mechanisms in humans in some cases. In the future, careful analysis of these tissues in mouse and zebrafish mutants that display scoliosis should be considered. Case in point, a recent report of a *wavy* mutant medaka, which displays characteristics of IS [145], may be due to dysplasia of the intervertebral disc region. This phenotype may be reminiscent of work showing that defective fusion of notochord vacuolated cells during larval development is sufficient to generate scoliosis and vertebral fusions in zebrafish [22]. It will be interesting to address these issues in mouse models as well as the functional role of the inner most cellular/tissue layer of the NP are likewise unknown (Fig. 5.2A', C). One could imagine that if these cells have a role in growth in homeostasis of the IVD, even subtle defects in their function could contribute to a loss of spine stability. In conclusion, clear structural differences in the mouse and zebrafish spine exist; undoubtedly caveats exist for all animal models of human disease. For this reason, the use of animal models to study IS should take care to reflect on anatomical differences and be cognizant of other potential caveats for each model system.

Considerations of Analysis of Spinal Curvatures in Mouse and Zebrafish

When working with mouse models, it is imperative to ensure careful alignment and placement of mice for X-ray analysis of the spine. In particular, we find that a "scoliosis" can be falsely observed in mice that are improperly placed in the X-ray scanner, and when readjusted, with hips and shoulders carefully aligned and adjusted along the AP axis, these same mice do not display spinal curvature (Fig. 5.3). This imaging artifact is likely due to the normal kyphosis of the mouse that is best seen on lateral views. When imaged appropriately using a lateral view, extreme kyphosis in mice may also be abnormal. Indeed, similar spinal pathology is observed in human scoliosis representing a spectrum of rotational deformities that often includes kyphosis (in combination, this deformity may be referred to as kyphoscoliosis). For example, humans with Marfan syndrome have increased incidence of both scoliotic and kyphotic defects of the spine. Mouse model of Marfan syndrome has often been described as having extreme kyphosis [146], with no description of scoliosis, although it is unclear whether scoliosis was even evaluated in this study.

Our personal experience with zebrafish models of IS has not uncovered this susceptibility to positioning affects. Indeed, mutant zebrafish with scoliosis are uniquely identified in swimming fish; however, we have observed some mutant

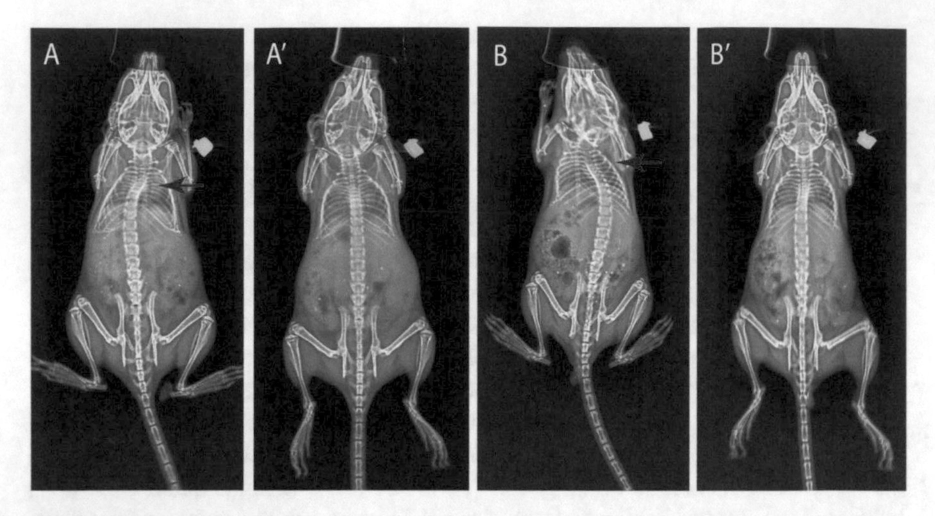

Fig. 5.3 Positioning of mice is critical for accurate determination of spine curvatures. Two wild-type C57BL/6J(JAX) mice at P40 imaged live under isoflurane anesthesia (**A** and **B** panels). In some cases, mice that imaged without proper care for placement will show an obvious spine curvature (red arrows) (**A**, **B**). However, after manipulation and lengthning of the spine and alignment of the shoulders and hips by gentle traction with the thumb and fingers you can balance the mouse on its ribcage, which allow for imageing of normal spine alignment (**A'**, **B'**).

zebrafish that display very mild scoliosis that are not readily noticeable without X-ray or skeletal preparation analysis (unpublished observations). This is in stark contrast to our observations in mouse where even severe thoracic scoliosis is not readily noticeable with X-ray analysis. Importantly this suggests that many more mouse mutant models may exist that have never been reported to have spine defects simply due to lack of observation, this is recently been supported by clear indications of scoliosis from high-throughput analysis of over 3000 novel mouse mutant strains as part of the International Mouse Phenotyping Project [147, 148].

Conclusions

While no single animal model can replicate the pathophysiology of the human spine clearly, animal models including teleost and mouse can be useful in studying the molecular genetics and mechanics of spine development and disease. While genetic studies in human IS are certain to become more powerful ways of advancing its molecular genetics, we argue that validation of these findings in animal models is critical to ensuring actionable biomedical research whether in the model or in human-derived in vitro culture models. Moreover, the discovery of the cellular pathogenesis of IS in these animals is important for advancing approval for studies of resected tissue from human IS and nonpathogenic patient sources.

Moving forward, it is worthwhile to consider the caveats of each model and start to hold future study of IS models to a higher standard, for instance: (i) in particular for zebrafish, we suggest that analysis must be done in a relatively mature spine, in contrast making conclusions of relevance to IS based on kinking body plans in embryonic zebrafish; (ii) in particular for mouse, care should be taken to ensure accurate alignment of the spine prior to analysis of the spine curvature by X-ray as false positives are easily observed; (iii) we suggest levels of validity that might help to address the distinctive hallmarks of human IS for individual models; (iv) where possible we suggest that analysis of IS models should be done using reproducible stable transgenic or heritable genetic animals or highly reproducible mechanically induced models.

Acknowledgments The authors would like to acknowledge Roberto Gonzalez for zebrafish histology and Drs. Michel Bagnat, Christina Gurnett, and Gabriel Haller for critical discussion of this manuscript. This work was supported in part by the National Institute of Arthritis and Musculoskeletal and Skin Diseases (R01-AR072009).

References

1. Smith LJ, Nerurkar NL, Choi KS, Harfe BD, Elliott DM. Degeneration and regeneration of the intervertebral disc: lessons from development. Dis Model Mech. 2011;4(1):31–41.
2. Eckalbar WL, Fisher RE, Rawls A, Kusumi K. Scoliosis and segmentation defects of the vertebrae. Wiley Interdiscip Rev Dev Biol. 2012;1(3):401–23.
3. Koehl MA, Quillin KJ, Pell CA. Mechanical design of fiber-wound hydraulic skeletons: the stiffening and straightening of embryonic notochords. Am Zool. 2000;40:28–41.
4. Adams DS, Keller R, Koehl MA. The mechanics of notochord elongation, straightening and stiffening in the embryo of Xenopus laevis. Development. 1990;110(1):115–30.
5. Glickman NS, Kimmel CB, Jones MA, Adams RJ. Shaping the zebrafish notochord. Development. 2003;130(5):873–87.
6. Shapiro IM, Risbud MV. Introduction to the structure, function, and comparative anatomy of the vertebrae and the intervertebral disc. In: Shapiro IM, Risbud MV, editors. The intervertebral disc: molecular and structural studies of the disc in health and disease. Vienna: Springer; 2014. p. 3–15.
7. Choi KS, Cohn MJ, Harfe BD. Identification of nucleus pulposus precursor cells and notochordal remnants in the mouse: implications for disk degeneration and chordoma formation. Dev Dyn. 2008;237(12):3953–8.
8. Newton Ede MM, Jones SW. Adolescent idiopathic scoliosis: evidence for intrinsic factors driving aetiology and progression. Int Orthop. 2016;40(10):2075–80.
9. Jackson HC 2nd, Winkelmann RK, Bickel WH. Nerve endings in the human lumbar spinal column and related structures. J Bone Joint Surg Am. 1966;48(7):1272–81.
10. Kojima Y, Maeda T, Arai R, Shichikawa K. Nerve supply to the posterior longitudinal ligament and the intervertebral disc of the rat vertebral column as studied by acetylcholinesterase histochemistry. I. Distribution in the lumbar region. J Anat. 1990;169:237–46.
11. Man GC, Wang WW, Yim AP, Wong JH, Ng TB, Lam TP, et al. A review of pinealectomy-induced melatonin-deficient animal models for the study of etiopathogenesis of adolescent idiopathic scoliosis. Int J Mol Sci. 2014;15(9):16484–99.
12. Lombardi G, Akoume MY, Colombini A, Moreau A, Banfi G. Biochemistry of adolescent idiopathic scoliosis. Adv Clin Chem. 2011;54:165–82.

13. Normand E, Franco A, Moreau A, Marcil V. Dipeptidyl Peptidase-4 and adolescent idiopathic scoliosis: expression in osteoblasts. Sci Rep. 2017;7(1):3173.
14. Blecher R, Krief S, Galili T, Biton IE, Stern T, Assaraf E, et al. The proprioceptive system masterminds spinal alignment: insight into the mechanism of scoliosis. Dev Cell. 2017;42(4):388–99. e3
15. Pourquie O. Vertebrate segmentation: from cyclic gene networks to scoliosis. Cell. 2011;145(5):650–63.
16. Giampietro PF, Dunwoodie SL, Kusumi K, Pourquie O, Tassy O, Offiah AC, et al. Progress in the understanding of the genetic etiology of vertebral segmentation disorders in humans. Ann N Y Acad Sci. 2009;1151:38–67.
17. Sparrow DB, Chapman G, Dunwoodie SL. The mouse notches up another success: understanding the causes of human vertebral malformation. Mamm Genome. 2011;22(7–8):362–76.
18. Gansner JM, Mendelsohn BA, Hultman KA, Johnson SL, Gitlin JD. Essential role of lysyl oxidases in notochord development. Dev Biol. 2007;307(2):202–13.
19. Gansner JM, Gitlin JD. Essential role for the alpha 1 chain of type VIII collagen in zebrafish notochord formation. Dev Dyn. 2008;237(12):3715–26.
20. Christiansen HE, Lang MR, Pace JM, Parichy DM. Critical early roles for col27a1a and col27a1b in zebrafish notochord morphogenesis, vertebral mineralization and post-embryonic axial growth. PLoS One. 2009;4(12):e8481.
21. Gray RS, Wilm TP, Smith J, Bagnat M, Dale RM, Topczewski J, et al. Loss of col8a1a function during zebrafish embryogenesis results in congenital vertebral malformations. Dev Biol. 2014;386(1):72–85.
22. Ellis K, Bagwell J, Bagnat M. Notochord vacuoles are lysosome-related organelles that function in axis and spine morphogenesis. J Cell Biol. 2013;200(5):667–79.
23. Sparrow DB, Chapman G, Smith AJ, Mattar MZ, Major JA, O'Reilly VC, et al. A mechanism for gene-environment interaction in the etiology of congenital scoliosis. Cell. 2012;149(2):295–306.
24. Purkiss SB, Driscoll B, Cole WG, Alman B. Idiopathic scoliosis in families of children with congenital scoliosis. Clin Orthop Relat Res. 2002;401:27–31.
25. Guo L, Yamashita H, Kou I, Takimoto A, Meguro-Horike M, Horike S, et al. Functional investigation of a non-coding variant associated with adolescent idiopathic scoliosis in zebrafish: elevated expression of the ladybird Homeobox gene causes body Axis deformation. PLoS Genet. 2016;12(1):e1005802.
26. Hayes M, Gao X, Yu LX, Paria N, Henkelman RM, Wise CA, et al. ptk7 mutant zebrafish models of congenital and idiopathic scoliosis implicate dysregulated Wnt signalling in disease. Nat Commun. 2014;5:4777.
27. McGreevy JW, Hakim CH, McIntosh MA, Duan D. Animal models of Duchenne muscular dystrophy: from basic mechanisms to gene therapy. Dis Model Mech. 2015;8(3):195–213.
28. Harrison DJ, Webb PJ. Scoliosis in the Rett syndrome: natural history and treatment. Brain and Development. 1990;12(1):154–6.
29. Taylor LJ. Severe spondylolisthesis and scoliosis in association with Marfan's syndrome. Case report and review of the literature. Clin Orthop Relat Res. 1987;221:207–11.
30. Shirley ED, Demaio M, Bodurtha J. Ehlers-danlos syndrome in orthopaedics: etiology, diagnosis, and treatment implications. Sports Health. 2012;4(5):394–403.
31. Blanco G, Coulton GR, Biggin A, Grainge C, Moss J, Barrett M, et al. The kyphoscoliosis (ky) mouse is deficient in hypertrophic responses and is caused by a mutation in a novel muscle-specific protein. Hum Mol Genet. 2001;10(1):9–16.
32. Chen F, Guo R, Itoh S, Moreno L, Rosenthal E, Zappitelli T, et al. First mouse model for combined osteogenesis imperfecta and Ehlers-Danlos syndrome. J Bone Miner Res. 2014;29(6):1412–23.
33. Haller G, Alvarado D, McCall K, Yang P, Cruchaga C, Harms M, et al. A polygenic burden of rare variants across extracellular matrix genes among individuals with adolescent idiopathic scoliosis. Hum Mol Genet. 2016;25(1):202–9.

34. Buchan JG, Alvarado DM, Haller GE, Cruchaga C, Harms MB, Zhang T, et al. Rare variants in FBN1 and FBN2 are associated with severe adolescent idiopathic scoliosis. Hum Mol Genet. 2014;23(19):5271–82.
35. Cheng JC, Castelein RM, Chu WC, Danielsson AJ, Dobbs MB, Grivas TB, et al. Adolescent idiopathic scoliosis. Nat Rev Dis Primers. 2015;1:15030.
36. van der Worp HB, Howells DW, Sena ES, Porritt MJ, Rewell S, O'Collins V, et al. Can animal models of disease reliably inform human studies? PLoS Med. 2010;7(3):e1000245.
37. McGonigle P, Ruggeri B. Animal models of human disease: challenges in enabling translation. Biochem Pharmacol. 2014;87(1):162–71.
38. Boszczyk BM, Boszczyk AA, Putz R. Comparative and functional anatomy of the mammalian lumbar spine. Anat Rec. 2001;264(2):157–68.
39. Langenskiold A, Michelsson JE. Experimental progressive scoliosis in the rabbit. J Bone Joint Surg Br. 1961;43-B:116–20.
40. Langenskiold A, Michelsson JE. Experimental scoliosis. Acta Orthop Scand. 1959;29:158–9.
41. Langenskiold A, Michelsson JE. The pathogenesis of experimental progressive scoliosis. Acta Orthop Scand Suppl. 1962;59:1–26.
42. Kubota K, Doi T, Murata M, Kobayakawa K, Matsumoto Y, Harimaya K, et al. Disturbance of rib cage development causes progressive thoracic scoliosis: the creation of a nonsurgical structural scoliosis model in mice. J Bone Joint Surg Am. 2013;95(18):e130.
43. Stokes IA, Laible JP. Three-dimensional osseo-ligamentous model of the thorax representing initiation of scoliosis by asymmetric growth. J Biomech. 1990;23(6):589–95.
44. Andriacchi T, Schultz A, Belytschko T, Galante J. A model for studies of mechanical interactions between the human spine and rib cage. J Biomech. 1974;7(6):497–507.
45. Grivas TB, Burwell RG, Purdue M, Webb JK, Moulton A. A segmental analysis of thoracic shape in chest radiographs of children. Changes related to spinal level, age, sex, side and significance for lung growth and scoliosis. J Anat. 1991;178:21–38.
46. Gurnett CA, Alaee F, Bowcock A, Kruse L, Lenke LG, Bridwell KH, et al. Genetic linkage localizes an adolescent idiopathic scoliosis and pectus excavatum gene to chromosome 18 q. Spine. 2009;34(2):E94–100.
47. Hong JY, Suh SW, Park HJ, Kim YH, Park JH, Park SY. Correlations of adolescent idiopathic scoliosis and pectus excavatum. J Pediatr Orthop. 2011;31(8):870–4.
48. Dubousset J, Wicart P, Pomero V, Barois A, Estournet B. Spinal penetration index: new three-dimensional quantified reference for lordoscoliosis and other spinal deformities. J Orthop Sci. 2003;8(1):41–9.
49. Doi T, Harimaya K, Matsumoto Y, Iwamoto Y. Aortic location and flat chest in scoliosis: a prospective study. Fukuoka Igaku Zasshi. 2011;102(1):14–9.
50. Ilharreborde B, Dubousset J, Le Huec JC. Use of EOS imaging for the assessment of scoliosis deformities: application to postoperative 3D quantitative analysis of the trunk. Eur Spine J. 2014;23(Suppl 4):S397–405.
51. Caballero A, Barrios C, Burgos J, Hevia E, Correa C. Vertebral growth modulation by hemicircumferential electrocoagulation: an experimental study in pigs. Eur Spine J. 2011;20(Suppl 3):367–75.
52. Catanzariti JF, Agnani O, Guyot MA, Wlodyka-Demaille S, Khenioui H, Donze C. Does adolescent idiopathic scoliosis relate to vestibular disorders? A systematic review. Ann Phys Rehabil Med. 2014;57(6–7):465–79.
53. Hawasli AH, Hullar TE, Dorward IG. Idiopathic scoliosis and the vestibular system. Eur Spine J. 2015;24(2):227–33.
54. Hitier M, Hamon M, Denise P, Lacoudre J, Thenint MA, Mallet JF, et al. Lateral Semicircular Canal asymmetry in idiopathic scoliosis: an early link between biomechanical, hormonal and neurosensory theories? PLoS One. 2015;10(7):e0131120.
55. Noshchenko A, Hoffecker L, Lindley EM, Burger EL, Cain CM, Patel VV, et al. Predictors of spine deformity progression in adolescent idiopathic scoliosis: a systematic review with meta-analysis. World J Orthop. 2015;6(7):537–58.

56. Pialasse JP, Mercier P, Descarreaux M, Simoneau M. Sensorimotor control impairment in young adults with idiopathic scoliosis compared with healthy controls. J Manip Physiol Ther. 2016;39(7):473–9.
57. Lambert FM, Malinvaud D, Glaunes J, Bergot C, Straka H, Vidal PP. Vestibular asymmetry as the cause of idiopathic scoliosis: a possible answer from Xenopus. J Neurosci. 2009;29(40):12477–83.
58. Lambert FM, Malinvaud D, Gratacap M, Straka H, Vidal PP. Restricted neural plasticity in vestibulospinal pathways after unilateral labyrinthectomy as the origin for scoliotic deformations. J Neurosci. 2013;33(16):6845–56.
59. Dahlhoff M, Emrich D, Wolf E, Schneider MR. Increased activation of the epidermal growth factor receptor in transgenic mice overexpressing epigen causes peripheral neuropathy. Biochim Biophys Acta. 2013;1832(12):2068–76.
60. Smit JJ, Baas F, Hoogendijk JE, Jansen GH, van der Valk MA, Schinkel AH, et al. Peripheral neuropathy in mice transgenic for a human MDR3 P-glycoprotein mini-gene. J Neurosci. 1996;16(20):6386–93.
61. Mogha A, Benesh AE, Patra C, Engel FB, Schoneberg T, Liebscher I, et al. Gpr126 functions in Schwann cells to control differentiation and myelination via G-protein activation. J Neurosci. 2013;33(46):17976–85.
62. Karner CM, Long F, Solnica-Krezel L, Monk KR, Gray RS. Gpr126/Adgrg6 deletion in cartilage models idiopathic scoliosis and pectus excavatum in mice. Hum Mol Genet. 2015;24(15):4365–73.
63. Buchan JG, Gray RS, Gansner JM, Alvarado DM, Burgert L, Gitlin JD, et al. Kinesin family member 6 (kif6) is necessary for spine development in zebrafish. Dev Dyn. 2014;243(12):1646–57.
64. Boswell CW, Ciruna B. Understanding idiopathic scoliosis: a new zebrafish School of Thought. Trends Genet. 2017;33(3):183–96.
65. Yang Z, Xie Y, Chen J, Zhang D, Yang C, Li M. High selenium may be a risk factor of adolescent idiopathic scoliosis. Med Hypotheses. 2010;75(1):126–7.
66. Lloyd HMS, Kirchhoff CA. Case study: scoliosis in a bonobo (Pan paniscus). J Med Primatol. 2018 Apr;47(2):114–116. doi: 10.1111/jmp.12325. Epub 2017 Nov 29.
67. Naique SB, Porter R, Cunningham AA, Hughes SP, Sanghera B, Amis AA. Scoliosis in an orangutan. Spine. 2003;28(7):E143–5.
68. Berghan J, VIsser IN. Vertebral column malformations in New Zealand delphinids with a review of cases world wide. Aquat Mamm. 2000;26(1):17–25.
69. Ambert AM, Samuelson MM, Pitchford JL, Solangi M. Visually detectable vertebral malformations of a bottlenose dolphin (Tursiops truncatus) in the Mississippi sound. Aquat Mamm. 2017;43(4):6.
70. Andrews B, Davis W, Parham D. Corporate response and facilitation of the rehabilitation of a California gray whale calf. Acad Radiol. 2001;Aquatic Mammals 273:209–11.
71. Ellis Giddens W, Ryland M, Casson CJ. Idiopathic scoliosis in a Newborn Sea otter, Enhydra lutris (L.). J Wildl Dis. 1984;20(3):248–50.
72. Mochida J, Benson DR, Abbott U, Rucker RB. Neuromorphometric changes in the ventral spinal roots in a scoliotic animal. Spine. 1993;18(3):350–5.
73. Nakai S. Histological and histochemical changes in the neck muscles of spontaneously occurring scoliosis in a special strain of Japanese quail, SQOHM. Nihon Seikeigeka Gakkai Zasshi. 1990;64(4):229–39.
74. Sobajima S, Kin A, Baba I, Kanbara K, Semoto Y, Abe M. Implication for melatonin and its receptor in the spinal deformities of hereditary Lordoscoliotic rabbits. Spine. 2003;28(6):554–8.
75. Grimes DT, Boswell CW, Morante NF, Henkelman RM, Burdine RD, Ciruna B. Zebrafish models of idiopathic scoliosis link cerebrospinal fluid flow defects to spine curvature. Science. 2016;352(6291):1341–4.

76. Gorman KF, Tredwell SJ, Breden F. The mutant guppy syndrome curveback as a model for human heritable spinal curvature. Spine. 2007;32(7):735–41.
77. Gorman KF, Christians JK, Parent J, Ahmadi R, Weigel D, Dreyer C, et al. A major QTL controls susceptibility to spinal curvature in the curveback guppy. BMC Genet. 2011;12(1):16.
78. Lieschke GJ, Currie PD. Animal models of human disease: zebrafish swim into view. Nat Rev Genet. 2007;8(5):353–67.
79. Hwang WY, Fu Y, Reyon D, Maeder ML, Kaini P, Sander JD, et al. Heritable and precise zebrafish genome editing using a CRISPR-Cas system. PLoS One. 2013;8(7):e68708.
80. Auer TO, Duroure K, De Cian A, Concordet JP, Del Bene F. Highly efficient CRISPR/Cas9-mediated knock-in in zebrafish by homology-independent DNA repair. Genome Res. 2014;24(1):142–53.
81. Luderman LN, Unlu G, Knapik EW. Zebrafish developmental models of skeletal diseases. Curr Top Dev Biol. 2017;124:81–124.
82. Fisher S, Jagadeeswaran P, Halpern ME. Radiographic analysis of zebrafish skeletal defects. Dev Biol. 2003;264(1):64–76.
83. Henke K, Daane JM, Hawkins MB, Dooley CM, Busch-Nentwich EM, Stemple DL, et al. Genetic screen for postembryonic development in the zebrafish (Danio rerio): dominant mutations affecting adult form. Genetics. 2017;207(2):609–23.
84. Paul S, Schindler S, Giovannone D, de Millo Terrazzani A, Mariani FV, Crump JG. Ihha induces hybrid cartilage-bone cells during zebrafish jawbone regeneration. Development. 2016;143(12):2066–76.
85. Huitema LF, Apschner A, Logister I, Spoorendonk KM, Bussmann J, Hammond CL, et al. Entpd5 is essential for skeletal mineralization and regulates phosphate homeostasis in zebrafish. Proc Natl Acad Sci U S A. 2012;109(52):21372–7.
86. Mackay EW, Apschner A, Schulte-Merker S. Vitamin K reduces hypermineralisation in zebrafish models of PXE and GACI. Development. 2015;142(6):1095–101.
87. Kou I, Takahashi Y, Johnson TA, Takahashi A, Guo L, Dai J, et al. Genetic variants in GPR126 are associated with adolescent idiopathic scoliosis. Nat Genet. 2013;45(6):676–9.
88. Monk KR, Naylor SG, Glenn TD, Mercurio S, Perlin JR, Dominguez C, et al. A G protein-coupled receptor is essential for Schwann cells to initiate myelination. Science. 2009;325(5946):1402–5.
89. Geng FS, Abbas L, Baxendale S, Holdsworth CJ, Swanson AG, Slanchev K, et al. Semicircular canal morphogenesis in the zebrafish inner ear requires the function of gpr126 (lauscher), an adhesion class G protein-coupled receptor gene. Development. 2013;140(21):4362–74.
90. Takahashi Y, Kou I, Takahashi A, Johnson TA, Kono K, Kawakami N, et al. A genome-wide association study identifies common variants near LBX1 associated with adolescent idiopathic scoliosis. Nat Genet. 2011;43(12):1237–40.
91. Cao Y, Min J, Zhang Q, Li H, Li H. Associations of LBX1 gene and adolescent idiopathic scoliosis susceptibility: a meta-analysis based on 34,626 subjects. BMC Musculoskelet Disord. 2016;17:309.
92. Chettier R, Nelson L, Ogilvie JW, Albertsen HM, Ward K. Haplotypes at LBX1 have distinct inheritance patterns with opposite effects in adolescent idiopathic scoliosis. PLoS One. 2015;10(2):e0117708.
93. Londono D, Kou I, Johnson TA, Sharma S, Ogura Y, Tsunoda T, et al. A meta-analysis identifies adolescent idiopathic scoliosis association with LBX1 locus in multiple ethnic groups. J Med Genet. 2014;51(6):401–6.
94. Brohmann H, Jagla K, Birchmeier C. The role of Lbx1 in migration of muscle precursor cells. Development. 2000;127(2):437–45.
95. Kruger M, Schafer K, Braun T. The homeobox containing gene Lbx1 is required for correct dorsal-ventral patterning of the neural tube. J Neurochem. 2002;82(4):774–82.
96. Muller T, Brohmann H, Pierani A, Heppenstall PA, Lewin GR, Jessell TM, et al. The homeodomain factor lbx1 distinguishes two major programs of neuronal differentiation in the dorsal spinal cord. Neuron. 2002;34(4):551–62.

97. Sieber MA, Storm R, Martinez-de-la-Torre M, Muller T, Wende H, Reuter K, et al. Lbx1 acts as a selector gene in the fate determination of somatosensory and viscerosensory relay neurons in the hindbrain. J Neurosci. 2007;27(18):4902–9.
98. Kilian B, Mansukoski H, Barbosa FC, Ulrich F, Tada M, Heisenberg CP. The role of Ppt/Wnt5 in regulating cell shape and movement during zebrafish gastrulation. Mech Dev. 2003;120(4):467–76.
99. Madsen EC, Gitlin JD. Zebrafish mutants calamity and catastrophe define critical pathways of gene-nutrient interactions in developmental copper metabolism. PLoS Genet. 2008;4(11):e1000261.
100. Mendelsohn BA, Yin C, Johnson SL, Wilm TP, Solnica-Krezel L, Gitlin JD. Atp7a determines a hierarchy of copper metabolism essential for notochord development. Cell Metab. 2006;4(2):155–62.
101. Hubaud A, Pourquie O. Signalling dynamics in vertebrate segmentation. Nat Rev Mol Cell Biol. 2014;15(11):709–21.
102. Yu X, Ng CP, Habacher H, Roy S. Foxj1 transcription factors are master regulators of the motile ciliogenic program. Nat Genet. 2008;40(12):1445–53.
103. Parichy DM, Elizondo MR, Mills MG, Gordon TN, Engeszer RE. Normal table of postembryonic zebrafish development: staging by externally visible anatomy of the living fish. Dev Dyn. 2009;238(12):2975–3015.
104. Turgut M, Cullu E, Uysal A, Yurtseven ME, Alparslan B. Chronic changes in cerebrospinal fluid pathways produced by subarachnoid kaolin injection and experimental spinal cord trauma in the rabbit: their relationship with the development of spinal deformity. An electron microscopic study and magnetic resonance imaging evaluation. Neurosurg Rev. 2005;28(4):289–97.
105. Chuma A, Kitahara H, Minami S, Goto S, Takaso M, Moriya H. Structural scoliosis model in dogs with experimentally induced syringomyelia. Spine. 1997;22(6):589–94. discussion 95
106. Godzik J, Dardas A, Kelly MP, Holekamp TF, Lenke LG, Smyth MD, et al. Comparison of spinal deformity in children with Chiari I malformation with and without syringomyelia: matched cohort study. Eur Spine J. 2016;25(2):619–26.
107. Ogura Y, Kou I, Takahashi Y, Takeda K, Minami S, Kawakami N, et al. A functional variant in MIR4300HG, the host gene of microRNA MIR4300 is associated with progression of adolescent idiopathic scoliosis. Hum Mol Genet. 2017;26(20):4086–92.
108. Patten SA, Margaritte-Jeannin P, Bernard JC, Alix E, Labalme A, Besson A, et al. Functional variants of POC5 identified in patients with idiopathic scoliosis. J Clin Invest. 2015;125(3):1124–8.
109. Becker-Heck A, Zohn IE, Okabe N, Pollock A, Lenhart KB, Sullivan-Brown J, et al. The coiled-coil domain containing protein CCDC40 is essential for motile cilia function and left-right axis formation. Nat Genet. 2011;43(1):79–84.
110. Jaffe KM, Grimes DT, Schottenfeld-Roames J, Werner ME, Ku TS, Kim SK, et al. c21orf59/kurly controls both cilia motility and polarization. Cell Rep. 2016;14(8):1841–9.
111. Serluca FC, Xu B, Okabe N, Baker K, Lin SY, Sullivan-Brown J, et al. Mutations in zebrafish leucine-rich repeat-containing six-like affect cilia motility and result in pronephric cysts, but have variable effects on left-right patterning. Development. 2009;136(10):1621–31.
112. Sullivan-Brown J, Schottenfeld J, Okabe N, Hostetter CL, Serluca FC, Thiberge SY, et al. Zebrafish mutations affecting cilia motility share similar cystic phenotypes and suggest a mechanism of cyst formation that differs from pkd2 morphants. Dev Biol. 2008;314(2):261–75.
113. Eisen JS, Smith JC. Controlling morpholino experiments: don't stop making antisense. Development. 2008;135(10):1735–43.
114. Kim HK, Aruwajoye O, Sucato D, Richards BS, Feng GS, Chen D, et al. Induction of SHP2 deficiency in chondrocytes causes severe scoliosis and kyphosis in mice. Spine. 2013;38(21):E1307–12.
115. Tidyman WE, Rauen KA. The RASopathies: developmental syndromes of Ras/MAPK pathway dysregulation. Curr Opin Genet Dev. 2009;19(3):230–6.

116. Henry SP, Liang S, Akdemir KC, de Crombrugghe B. The postnatal role of Sox9 in cartilage. J Bone Miner Res. 2012;27(12):2511–25.
117. Akiyama H, Chaboissier MC, Martin JF, Schedl A, de Crombrugghe B. The transcription factor Sox9 has essential roles in successive steps of the chondrocyte differentiation pathway and is required for expression of Sox5 and Sox6. Genes Dev. 2002;16(21):2813–28.
118. Settle SH Jr, Rountree RB, Sinha A, Thacker A, Higgins K, Kingsley DM. Multiple joint and skeletal patterning defects caused by single and double mutations in the mouse Gdf6 and Gdf5 genes. Dev Biol. 2003;254(1):116–30.
119. Hartmann C, Tabin CJ. Wnt-14 plays a pivotal role in inducing synovial joint formation in the developing appendicular skeleton. Cell. 2001;104(3):341–51.
120. Lee WT, Cheung CS, Tse YK, Guo X, Qin L, Lam TP, et al. Association of osteopenia with curve severity in adolescent idiopathic scoliosis: a study of 919 girls. Osteoporos Int: a journal established as result of cooperation between the European Foundation for Osteoporosis and the National Osteoporosis Foundation of the USA. 2005;16(12):1924–32.
121. Hung VW, Qin L, Cheung CS, Lam TP, Ng BK, Tse YK, et al. Osteopenia: a new prognostic factor of curve progression in adolescent idiopathic scoliosis. J Bone Joint Surg Am. 2005;87(12):2709–16.
122. Colvin JS, Bohne BA, Harding GW, McEwen DG, Ornitz DM. Skeletal overgrowth and deafness in mice lacking fibroblast growth factor receptor 3. Nat Genet. 1996;12(4):390–7.
123. Deng C, Wynshaw-Boris A, Zhou F, Kuo A, Leder P. Fibroblast growth factor receptor 3 is a negative regulator of bone growth. Cell. 1996;84(6):911–21.
124. Valverde-Franco G, Liu H, Davidson D, Chai S, Valderrama-Carvajal H, Goltzman D, et al. Defective bone mineralization and osteopenia in young adult FGFR3−/− mice. Hum Mol Genet. 2004;13(3):271–84.
125. Valverde-Franco G, Binette JS, Li W, Wang H, Chai S, Laflamme F, et al. Defects in articular cartilage metabolism and early arthritis in fibroblast growth factor receptor 3 deficient mice. Hum Mol Genet. 2006;15(11):1783–92.
126. Gao C, Chen BP, Sullivan MB, Hui J, Ouellet JA, Henderson JE, et al. Micro CT analysis of spine architecture in a mouse model of scoliosis. Front Endocrinol (Lausanne). 2015;6:38.
127. Clin J, Aubin CE, Parent S, Labelle H. A biomechanical study of the Charleston brace for the treatment of scoliosis. Spine. 2010;35(19):E940–7.
128. MacIntyre NJ, Recknor CP, Grant SL, Recknor JC. Scores on the safe functional motion test predict incident vertebral compression fracture. Osteoporos Int : a journal established as result of cooperation between the European Foundation for Osteoporosis and the National Osteoporosis Foundation of the USA 2014;25(2):543–50.
129. Makrythanasis P, Temtamy S, Aglan MS, Otaify GA, Hamamy H, Antonarakis SE. A novel homozygous mutation in FGFR3 causes tall stature, severe lateral tibial deviation, scoliosis, hearing impairment, camptodactyly, and arachnodactyly. Hum Mutat. 2014;35(8):959–63.
130. Komatsu Y, Chusho H, Tamura N, Yasoda A, Miyazawa T, Suda M, et al. Significance of C-type natriuretic peptide (CNP) in endochondral ossification: analysis of CNP knockout mice. J Bone Miner Metab. 2002;20(6):331–6.
131. Tsuji T, Kunieda T. A loss-of-function mutation in natriuretic peptide receptor 2 (Npr2) gene is responsible for disproportionate dwarfism in cn/cn mouse. J Biol Chem. 2005;280(14):14288–92.
132. Miura K, Kim OH, Lee HR, Namba N, Michigami T, Yoo WJ, et al. Overgrowth syndrome associated with a gain-of-function mutation of the natriuretic peptide receptor 2 (NPR2) gene. Am J Med Genet A. 2014;164A(1):156–63.
133. Miura K, Namba N, Fujiwara M, Ohata Y, Ishida H, Kitaoka T, et al. An overgrowth disorder associated with excessive production of cGMP due to a gain-of-function mutation of the natriuretic peptide receptor 2 gene. PLoS One. 2012;7(8):e42180.
134. Waller-Evans H, Promel S, Langenhan T, Dixon J, Zahn D, Colledge WH, et al. The orphan adhesion-GPCR GPR126 is required for embryonic development in the mouse. PLoS One. 2010;5(11):e14047.

135. Monk KR, Oshima K, Jors S, Heller S, Talbot WS. Gpr126 is essential for peripheral nerve development and myelination in mammals. Development. 2011;138(13):2673–80.
136. Brochhausen C, Turial S, Muller FK, Schmitt VH, Coerdt W, Wihlm JM, et al. Pectus excavatum: history, hypotheses and treatment options. Interact Cardiovasc Thorac Surg. 2012;14(6):801–6.
137. Wu S, Sun X, Zhu W, Huang Y, Mou L, Liu M, et al. Evidence for GAL3ST4 mutation as the potential cause of pectus excavatum. Cell Res. 2012;22(12):1712–5.
138. Lefebvre V, Bhattaram P. Vertebrate skeletogenesis. Curr Top Dev Biol. 2010;90:291–317.
139. Fleming A, Keynes R, Tannahill D. A central role for the notochord in vertebral patterning. Development. 2004;131(4):873–80.
140. Grotmol S, Kryvi H, Nordvik K, Totland GK. Notochord segmentation may lay down the pathway for the development of the vertebral bodies in the Atlantic salmon. Anat Embryol (Berl). 2003;207(4–5):263–72.
141. Haga Y, Dominique VJ 3rd, Du SJ. Analyzing notochord segmentation and intervertebral disc formation using the twhh: gfp transgenic zebrafish model. Transgenic Res. 2009;18(5):669–83.
142. Cortes DH, Elliott DM. The intervertebral disc: overview of disc mechanics. In: Shapiro IM, Risbud MV, editors. The intervertebral disc: molecular and structural studies of the disc in health and disease. Vienna: Springer; 2014. p. 17–31.
143. Bruggeman BJ, Maier JA, Mohiuddin YS, Powers R, Lo Y, Guimaraes-Camboa N, et al. Avian intervertebral disc arises from rostral sclerotome and lacks a nucleus pulposus: implications for evolution of the vertebrate disc. Dev Dyn. 2012;241(4):675–83.
144. Inohaya K, Takano Y, Kudo A. The teleost intervertebral region acts as a growth center of the centrum: in vivo visualization of osteoblasts and their progenitors in transgenic fish. Dev Dyn. 2007;236(11):3031–46.
145. Irie K, Kuroda Y, Mimori N, Hayashi S, Abe M, Tsuji N, et al. Histopathology of a wavy medaka. J Toxicol Pathol. 2016;29(2):115–8.
146. Pereira L, Lee SY, Gayraud B, Andrikopoulos K, Shapiro SD, Bunton T, et al. Pathogenetic sequence for aneurysm revealed in mice underexpressing fibrillin-1. Proc Natl Acad Sci U S A. 1999;96(7):3819–23.
147. Dickinson ME, Flenniken AM, Ji X, Teboul L, Wong MD, White JK, et al. High-throughput discovery of novel developmental phenotypes. Nature. 2016;537(7621):508–14.
148. Meehan TF, Conte N, West DB, Jacobsen JO, Mason J, Warren J, et al. Disease model discovery from 3,328 gene knockouts by the international mouse phenotyping consortium. Nat Genet. 2017;49(8):1231–8.

Chapter 6
Current Understanding of Genetic Factors in Idiopathic Scoliosis

Carol A. Wise and Shiro Ikegawa

Introduction

Art and literature dating back to the beginnings of modern history have depicted the human struggle with scoliosis.

Perhaps the most notorious case of scoliosis in Western history was that of King Richard III of England, who Thomas More described as having "…croke backed, his left shoulder much higher than his right…." This image was borne out sensationally 527 years later when a twisted skeleton was unearthed and proven by DNA evidence to be almost certainly that of the fifteenth-century British monarch [1]. Subsequent analyses suggest that Richard III suffered from adolescent "idiopathic" scoliosis (AIS), so-called because it arises in children entering the adolescent growth spurt who appear to be otherwise healthy (Fig. 6.1a).

AIS is a puzzling disease that has evaded biologic understanding for decades despite numerous investigations [2]. Epidemiologic studies (described in more detail in Chap. 7) consistently identify high heritability and striking sexual dimorphism in AIS, with girls having a more than fivefold greater risk of progressive deformity than boys [3]. These observations support a disease model in which susceptibility to AIS is primarily driven by heritable factors that may differ between the sexes. In this model, the total mutational burden required to "activate" progressive AIS in males will be greater than that in females, a phenomenon known as the

C. A. Wise (✉)
Sarah M. and Charles E. Seay Center for Musculoskeletal Research, Texas Scottish Rite Hospital for Children, Dallas, TX, USA

Departments of Orthopaedic Surgery, Pediatrics, and McDermott Center for Human Growth and Development, University of Texas Southwestern Medical Center, Dallas, TX, USA
e-mail: carol.wise@tsrh.org

S. Ikegawa
Laboratory for Bone and Joint Diseases, RIKEN Center for Integrative Medical Sciences, Tokyo, Japan

K. Kusumi, S. L. Dunwoodie (eds.), *The Genetics and Development of Scoliosis*, https://doi.org/10.1007/978-3-319-90149-7_6

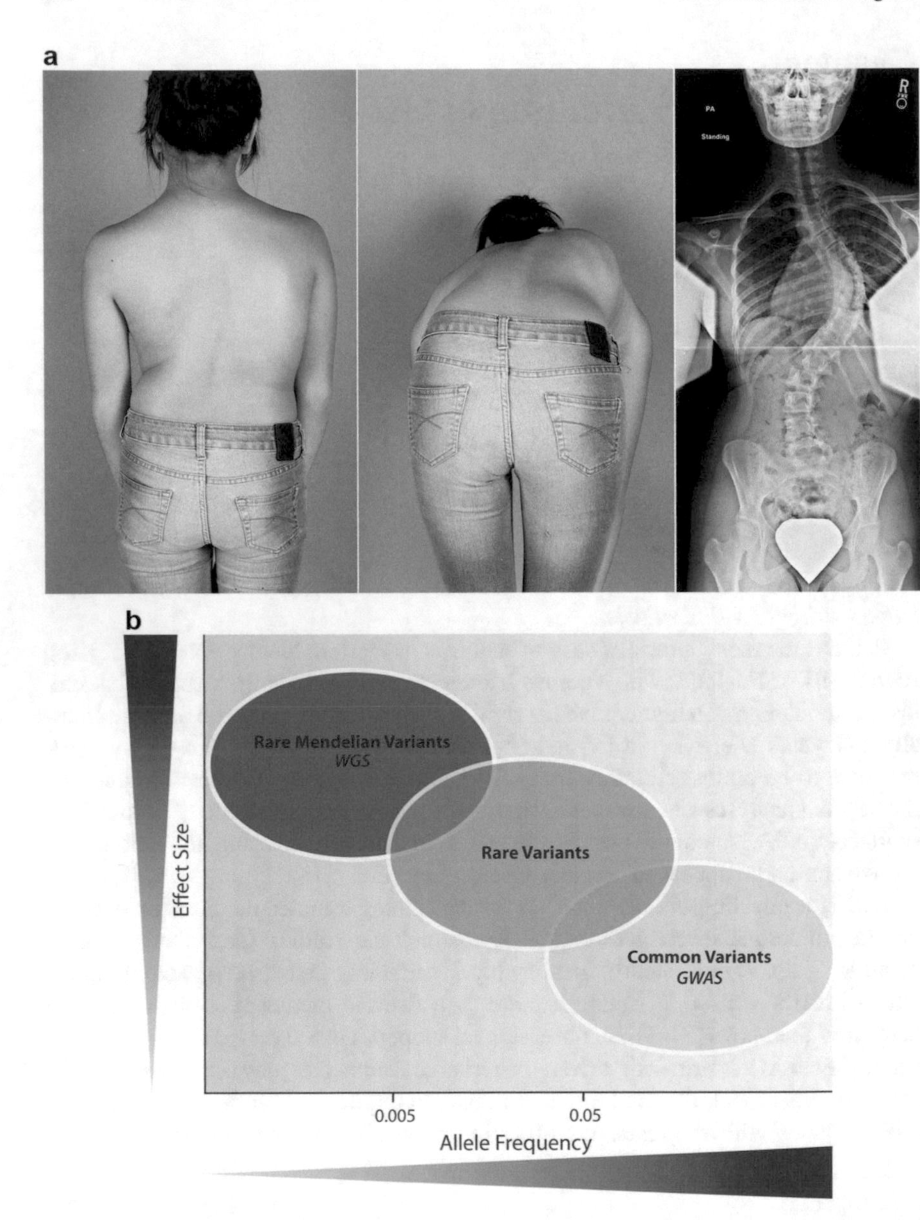

Fig. 6.1 (**a**) AIS in an adolescent female. Shoulder imbalance and rib hump are noted. Standing posteroanterior radiograph reveals a typical right thoracic curve with no visible anomalies. (**b**) Genetic heterogeneity in AIS. As depicted, the frequency of an allele (e.g., of a SNP) is expected to be inversely related to its effect size

Carter effect [4, 5]. Males with progressive AIS therefore should be fruitful for discovering deleterious AIS-causing mutations. The genetic underpinnings in AIS set up the exciting opportunity for mutation discovery in human populations that can get to biologic causality. Advantages to such human-based studies are that [1] animal models of AIS are few and may not fully recapitulate the phenotype; [2] the ability to completely and systematically survey each genome in the current era of high-throughput genomics can yield comprehensive results; and [3] these discoveries in sum may explain a significant fraction of overall disease risk.

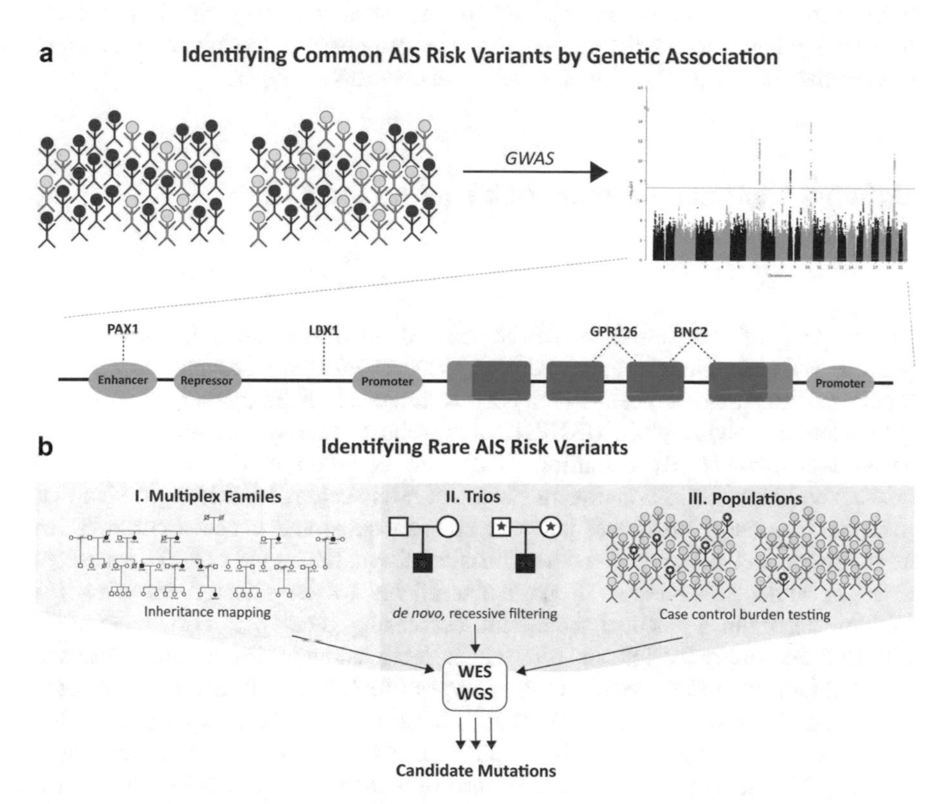

Fig. 6.2 (**a**) Framework for discovering common AIS genetic risk variants. Stick figures depict case and control populations, where cases are enriched for a particular common allele. After statistical comparisons the data are displayed as a Manhattan plot, with SNPs in chromosome order on the X-axis versus their evidence for association as measured by the (inverse $\log_{10}$) of the P-value. A representative gene is shown at the bottom, with exons shown as blue boxes, 5′ and 3′ untranslated regions shown in green, and noncoding regulatory elements shown as yellow ovals. For simplicity, we show replicated associations with AIS susceptibility by their relative location to the gene. So, for example, the chromosome 20p11.12 susceptibility locus maps within an *PAX1* enhancer; the chr10q24.31 signal is near the LBX1 promoter, etc. (**b**) Framework for discovering rare AIS genetic risk variants. The search space for rare disease variants detected by WES or WGS may be reduced to those co-segregating with disease in an extended family (left), those satisfying de novo or recessive inheritance models (middle), or by gene-based burden testing in cases and controls

As with other complex genetic disorders, we expect many heritable mutations with varying effect sizes and frequencies will comprise the total mutational burden in AIS. As depicted in Fig. 6.2, variants that are rare in a population tend to be more penetrant and confer a greater disease risk in an individual, whereas common variants are expected to be less penetrant and to confer relatively smaller effects [6]. In this chapter, we review progress in population-based gene discovery for AIS for solving the complex genetic architecture of AIS. For a review of family-based AIS gene discovery, please see Chap. 7. For simplicity, we have divided this review into discussions of "common" and "rare" variants (i.e., genetic mutations). The implications to epigenetics and systems biology are also discussed along with the prospects for using genetics to predict the likelihood of progression, to estimate individual disease risk, and as a clinical tool to delineate new AIS subtypes.

Defining Common Genetic Risk Factors for AIS

GWAS

The majority of genetic risk in AIS has been defined thus far by hypothesis-free genome-wide association studies (GWAS). Such studies are designed to map the location of genetic risk factors in the genome relative to fixed markers, usually single nucleotide polymorphisms (SNPs) that are robustly genotyped using microarray-based methods [7]. By definition, SNPs are common in the population and consequently can be informative in statistical tests such as logistic regression that measure frequency differences between affected cases and matched controls. An alternative statistical method, the transmission disequilibrium test (TDT), measures over- or under-transmission of particular alleles to the affected offspring [8] (Table 6.1). From a practical standpoint, decreasing genotyping costs have made GWAS increasingly feasible and affordable. Public and consortium initiatives such as the Database of Genotypes and Phenotypes (dbGaP) [9, 10] and the Wellcome Trust Case Control Consortium (WTCCC) have proven to be a major boon to the success of GWAS by providing controlled access to genotypes from other large studies [11]. Large-scale institutional efforts to biobank control DNA samples have likewise sped up discovery by GWAS and other platforms [12].

Further gains in power have been achieved by methods of genome-wide imputation, in which additional genotypes may be inferred by comparison to patterns of linkage disequilibrium (LD) in a reference dataset such as UK10K and 1000G [13] (Table 6.2). Imputing genotypes genome-wide can be computationally intense, but the method can pay off handsomely in terms of new discoveries that would have been missed otherwise [7, 14]. The post-genotyping phase of GWAS is also driven by a toolbox of publicly available resources. For example, SNP allele frequency data by population as provided in dbSNP and HapMap enable quality control checks to identify unmatched outlier samples. Other useful resources for interpreting genomic intervals underlying GWAS signals are given in Table 6.2, 6.3, and 6.4.

Table 6.1 Genetic definitions

Acronym	Name	Definition
GWAS	Genome-wide association study	An observational study of a genome-wide set of genetic variants in different individuals to see if any variant is associated with a trait
SNP	Single nucleotide polymorphism	A variation in a single nucleotide that occurs at a specific position in the genome, where each variation is present to some appreciable degree within a population (e.g., > 1%)
TDT	Transmission disequilibrium test	A family-based association test for the presence of genetic linkage between a genetic marker and a trait
LD	Linkage disequilibrium	The nonrandom association of alleles at different loci in a given population
eQTL	Expression quantitative trait locus	Genomic loci that contribute to variation in expression levels of mRNAs
SNV	Single nucleotide variant	A variation in a single nucleotide without any limitations of frequency and may arise in somatic cells
WES	Whole exome sequencing	Is a transcriptomics technique for sequencing all of the protein-coding genes in a genome (known as the exome)
WGS	Whole genome sequencing	The process of determining the complete DNA sequence of an organism's genome at a single time
RNA-seq	RNA sequencing	Uses next-generation sequencing (NGS) to reveal the presence and quantity of RNA in a biological sample at a given moment in time
ChIP-seq	Chromatin immunoprecipitation	A type of immunoprecipitation experimental technique used to investigate the interaction between proteins and DNA in the cell
ATAC-seq	Assay for transposase-accessible chromatin	A technique that identifies chromatin accessibility genome-wide
Hi-C	Extension of chromatin conformation capture (3C)	A technique that uses cross-linking and high-throughput sequencing to study long-range interactions genome-wide
Capture Hi-C	Modification of Hi-C	A technique that uses hybridization arrays to capture interactions with specific functional elements
ChIA-PET	Chromatin interaction analysis by paired-end tag sequencing	A technique that incorporates chromatin immunoprecipitation, ChIP-based enrichment, chromatin proximity ligation, paired-end tags, and high-throughput sequencing to determine long-range interactions genome-wide
Hi-ChIP	Extension of Hi-C	A technique designed to allow similar analysis as ChIA-PET with less input material
PLAC-seq	Proximity ligation-assisted ChIP-seq	A technique in which proximity ligation is conducted in nuclei prior to chromatin shearing and immunoprecipitation

Table 6.2 Public databases

Resource	Purpose	URL
dbGaP	Database of genotypes and phenotypes	https://www.ncbi.nlm.nih.gov/gap
WTCCC	Wellcome Trust case control consortium	https://www.wtccc.org.uk/
UK10K	United Kingdom ten thousand	https://www.uk10k.org/
1000G	1000 genomes	http://www.internationalgenome.org/
dbSNP	Database of single nucleotide polymorphisms	https://www.ncbi.nlm.nih.gov/snp/
NHGRI-EBI GWAS resource	Catalog of published genome-wide association studies	http://www.genome.gov/gwastudies/
gnomAD	Genome aggregation database	http://gnomad.broadinstitute.org/
ENCODE	Encyclopedia of DNA elements	https://www.encodeproject.org/
OMIM	Online Mendelian inheritance in man	https://www.omim.org/
GTEx	Genotype-tissue expression	https://www.gtexportal.org/home/

Table 6.3 Platforms and resources for population-based gene discovery

Technology	Readout	Controls	Statistical test	Clinical reference
SNP microarray genotyping	SNPs, CNVs	gnomAD, HapMap, dbGaP GWAS datasets	Logistic regression, trend tests, TDT	GWAS catalog
WES	SNVs, indels	gnomAD, EVS	Burden	ClinVar, OMIM
WGS	SNVs, indels, CNVs, SNPs	1000G, UK10K	Burden	ClinVar, OMIM, dbSNP

Because of the many tests performed in a typical GWAS (so-called multiple testing), a significance level of $P < 5 \times 10^{-8}$ is generally required for any association to be declared significant (GWAS-level significance). Follow-up studies to independently validate candidate SNPs can reveal false positives that may arise from random allele frequency differences or technical reasons and provide true effect sizes. Consequently, access to multiple large cohorts is needed as a starting point for any GWAS. It is important to note that while GWAS are proven to be powerful for mapping disease loci, genotyped SNPs may not be disease alleles themselves, and further fine-mapping approaches are needed to define causal mutations. The NHGRI-EBI Catalog of published genome-wide association studies aggregates published studies that can be queried by disease, chromosomal location, associated SNP, etc., allowing quick cross-study comparisons [15, 16].

Table 6.4 Susceptibility loci in AIS

Region	Marker	Phenotype	P-value (odds ratio)	Candidate gene	Reference
10q24.31	rs11190870	AIS susceptibility	1.24×10^{-19} (1.56)	*LBX1*	59
6q24.1	rs6570507	AIS susceptibility	1.27×10^{-14} (1.27)	*GPR126*	61
9p22.2	rs3904778	AIS susceptibility	2.46×10^{-13} (1.21)	*BNC2*	77
20p11.22	rs6137473	AIS susceptibility	2.15×10^{-10} (1.30)	*PAX1*	42
18q21.33	rs4940576	AIS susceptibility	2.22×10^{-12} (1.23)	*BCL2*	62
2q36.1	rs13398147	AIS susceptibility	7.59×10^{-13} (1.28)	*PAX3/EPHA4*	62
1p36.32	rs241215	AIS susceptibility	2.95×10^{-9} (0.83)	*AJAP1*	62
17q24.3	rs12946942	AIS curve severity	6×10^{-12} (1.56)	*SOX9, KCNJ2*	60
11q24.1	rs35333564	AIS curve severity	1.98×10^{-9} (1.56)	*MIR4300HG*	69

Regions in the human genome providing statistically significant evidence of harboring an AIS risk factor are denoted by their OMIM locus number (column one) where applicable. The markers shown are those that provided most significant evidence in the region

Here we summarize current progress in the discovery of both common and rare AIS risk variants, focusing on loci that have been identified by genome-wide methods and validated in multiple studies. For simplicity, we present each associated region by chromosome band and nearest gene, recognizing that some signals point to extragenic, noncoding disease loci.

Chr10q24.31: LBX1

The first definitive AIS locus was identified in a GWAS and follow-up replication in a Japanese population [17]. The most significantly associated SNP in the region was rs11190870 (P-value = 1.24×10^{-19}, OR = 1.56). This association was subsequently replicated in a Hong Kong Chinese population and cohorts of Han Chinese in Nanjing and Guangzhou [18, 19]. A more recent international meta-analysis using Japanese, Han Chinese, and Caucasian populations (5159 cases and 17,840 controls) supported rs11190870 as a global AIS susceptibility SNP ($P = 1.22 \times 10^{-45}$; OR = 1.60) [20]. rs11190870 is encoded within 80 kb of strong LD and very near the end of the *LBX1* gene that is encoded head-to-head with another gene *FLJ41350* (also called *LOC399806* or *LBX1-AS1* (for *LBX1* antisense RNA 1)).

LBX1 (OMIM 604255) is the vertebrate ortholog of the ladybird late (lbl) homeobox gene originally discovered in the fruit fly *Drosophila melanogaster* [21, 22]. In the fly, lbl participates in segmentation and cell specification of heart and muscle precursors. In the mouse *Lbx1* is specifically expressed during embryogenesis, with expression restricted to the developing central nervous system and muscles. *Lbx1* knockout (KO) mice had extensive muscle loss at birth [23–25]. The KO mice also had defects in heart looping and myocardial hyperplasia [26]. Lbx1 is suggested to control the expression of genes that guide lateral migration of muscle precursors and maintain their migratory potential. However, scoliosis was not described in the KO mice. Further, loss of *Lbx1* causes apoptotic loss of dorsal association neurons, cells of the dorsal spinal cord that relay somatosensory information, and disruption of dorsal horn innervation by nociceptive afferent neurons along with an increase in dorsal commissural neurons [27, 28]. Thus, *LBX1* may be implicated in myogenic and neurogenic etiology of AIS. *FLJ41350* is predicted to encode a 120 amino-acid human-specific protein. Its predicted protein has no known homology to other human proteins. Human *LBX1* has three zebrafish homologues, namely, lbx1a, lbx1b, and lbx2. It is interesting that both overexpression and morpholino knockdown of the three lbx genes produced zebrafish with scoliosis. Similar experiments with FLJ41350 did not yield scoliosis phenotypes. Thus, LBX1 itself may play a dosage-sensitive role in spine deformity [29].

Chr6q24: GPR126

In an expanded Japanese GWAS, significant associations were observed with SNPs on chromosome 6q24.1 within an intron of the *ADGRG6* gene, also referred to as *GPR126*. The association was replicated in Chinese and US-Caucasian (non-Hispanic white) populations (rs6570507 combined P-value = 1.27×10^{-14} (OR = 1.27)) and has since been replicated in multiple additional studies [30, 31]. Comparison to the catalog of published GWAS found that rs6570507 was also associated with trunk length in European populations [32]. *GPR126* (OMIM 612243) consists of 26 exons and encodes a 1250-aa adhesion G protein-coupled receptor. *GPR126* is highly expressed in human cartilage and in the proliferating cartilage of the mouse vertebral column, suggesting a role in spinal development [30]. In the ATDC5 cell line, *Gpr126* mRNA expression increases with early chondrogenic differentiation and is apparently regulated by the SOX9 transcription factor [33]. In the same cells, *Gpr126* overexpression increases expression of the cartilage marker genes *Col2a1* (encoding type II collagen) and *Acan* (encoding aggrecan), while its knockdown decreases their expression [34], indicating that Gpr126 is a positive regulator of cartilage differentiation.

GPR126 is a member of the adhesion G protein-coupled receptor (GPCR) family that functions in cell adhesion and migration [35]. In zebrafish, gpr126 is required for Schwann cell myelination, an effect that could be overcome by elevating cAMP levels with forskolin. Originally characterized as an orphan receptor, these data suggested that GPR126 signals through G proteins [36]. Mouse *Gpr126* null mutants fail to grow and also have severe neurologic deficits, displaying hypomyelinating peripheral neuropathy [37–39]. More recently, a novel genetic mouse model showing many of the hallmarks of AIS was provided by genetically removing *Gpr126* in osteoprogenitor cells. Axial cartilage cells lacking *Gpr126* become apoptotic prior to onset of AIS. This and additional experiments defined a critical role for Gpr126 during postembryonic cartilage development, including effects on the morphogenesis of the annulus fibrosis, chondrocyte survival, and the expression of *Gal3st4* [40].

Chr9p22.3–22.2: BNC2

Heritability studies (calculated as the square of the correlation coefficient, r^2) estimate that *LBX1*- and GPR126-associated loci only explain ~1% of the total genetic variance in AIS. To identify additional susceptibility gene(s) for AIS, a more powerful GWAS was designed by substantially increasing the number of the subjects to greater than fivefold and conducting a whole genome imputation and a meta-analysis. Analysis of the association of 4,420,789 SNPs for 2109 Japanese subjects with AIS and 11,140 control subjects yielded a third novel locus, rs3904778 on chromosome 9p22.2, that surpassed genome-wide significance and replicated in independent Japanese and Han Chinese populations (combined $P = 1.70 \times 10^{-13}$, OR = 1.21) [41]. The most significantly associated SNPs were in intron 3 of *BNC2* (OMIM 608669) that encodes basonuclin 2, a highly conserved transcription factor belonging to the *C2H2* group of zinc finger proteins. In humans, this gene is most highly expressed in the uterus and spinal cord, and expression is also evident in the bone and cartilage. Expression quantitative trait locus (eQTL) data in the GTEx database suggested that the associated alleles have the potential to upregulate *BNC2* transcription. *BNC2* overexpression produced body curvature and abnormal somite formation in zebrafish, suggesting that a gain of function increases AIS risk [41].

Chr20p11.22: PAX1

A GWAS in 3102 individuals of European descent (non-Hispanic white) identified significant associations with a locus on chromosome 20p11.22 that is distal to *PAX1*. This association was also found in independent North American, Japanese, and East Asian female cohorts [42]. Further investigation revealed that the association was driven almost entirely by females (combined rs6137473 $P = 2.15 \times 10^{-10}$, OR = 1.30) but not males (combined rs6137473 $P = 0.71$, OR = 1.08), identifying it as a sex-specific AIS locus. The 174-kb associated locus is distal to the *PAX1* (paired box 1, OMIM 167411) region, a transcription factor that is required for normal development of ventral vertebral structures [43]. To identify potential functional sequences, the region was annotated with predicted skeletal muscle regulatory elements as noted in the ENCODE database. One candidate sequence demonstrated enhancer activity in zebrafish somitic muscle and spinal cord, an activity that was abolished by AIS-associated SNPs. Thus, this sexually dimorphic AIS susceptibility locus was potentially refined to a noncoding enhancer that may regulate *PAX1* expression in spinal development.

Other Loci

A four-stage genome-wide association study (GWAS) of 4317 female Han Chinese AIS patients and 6016 controls detected the previously described association near *LBX1* and identified three novel susceptibility loci [44]. One locus on chromosome 1p36.32 represented by SNP rs241215 (P combined = 2.95×10^{-9}, OR = 0.83) is about 125 kb 5′ of the *AJAP1* gene at 2q36.1. *AJAP1* encodes adherens junction-associated protein 1, a transmembrane protein that interacts with E-cadherin/catenin complexes and may regulate cell adhesion, migration, and invasion [45]. The second locus represented by SNP rs13398147 (P combined = 7.59×10^{-13}, OR = 1.28) is in a region of about 500 kb between *PAX3* and *EPHA4*. The association is presumed to be driven by variation in noncoding regulatory elements with effects on *PAX3* or *EPHA4* or both. PAX3 regulates myogenesis in early development and neurogenesis in the neural tube [46–48]. In humans *PAX3* mutations are associated with syndromes involving hearing, limb, craniofacial, and pigmentation anomalies. In the mouse Pax3 together with Pax7 is necessary to maintain a source of myogenic cells that are required for skeletal muscle formation and maintenance after early embryonic stages [49]. *EPHA4* encodes a receptor protein tyrosine kinase that is critical for appropriate corticospinal tract development and axon guidance [50]. A third locus is on chromosome 18q21.33 represented by SNP rs4940576 (P combined = 2.22×10^{-12}, OR = 1.23) within the *BCL2* gene. The latter encodes the B cell lymphoma 2 proto-oncogene, an integral outer mitochondrial membrane protein that mediates resistance to apoptosis in a variety of cell types. In neurons BCL2 participates in neurogenesis during development and also enhances neuron survival throughout adult life [51]. In bone, *BCL2* overexpression can inhibit osteoblast differentiation [52]. No replication has been reported for these three new loci.

Association with Severity

One of the major goals in current clinical management of AIS is to accurately identify and aggressively treat those patients with a high likelihood of curve progression. Correlating factors include bone age at first presentation, sex, menarchal status, remaining growth, and curve pattern [2]. Genetic factors are presumed to influence curve progression, and targeted studies of a few SNPs in *ESR1*, *ESR2*, *MATN1*, and *IGF1* genes have suggested associations with AIS severity [53–56].

A comprehensive search for severity-correlated SNPs was conducted in a GWAS by using only severely affected AIS subjects, defined by having spinal deformity with Cobb angle above 40°. Through a two-stage association study using a total of ~12,000 Japanese subjects, six AIS-associated loci were identified that surpassed a genome-wide significance level [57]. Five of these were previously reported, three near *LBX1* on chromosome 10q24.31 and two on chromosome 6q24.1 in *GPR126*. The allele frequencies of SNPs in the *LBX1* and *GPR126* loci did not correlate with

AIS severity. One SNP, rs12946942 on chromosome 17q24.3, showed significant association in a recessive model ($P = 4.00 \times 10^{-8}$, OR = 2.05) and was replicated in a Han Chinese population (combined $P = 6.43 \times 10^{-12}$, OR = 2.21). This SNP also showed significant association between curve severity and genotype using the Kruskal-Wallis test [57]. rs12946942 and the LD block containing it are in a gene desert between *SOX9* and *KCNJ2*. Both of them are ~1 Mb away from the SNP but are excellent candidate genes.

SOX9 is a master transcriptional regulator of chondrogenesis and cartilage formation [58]. Recently the SOX5/6/9 trio proteins were shown to act cooperatively genome-wide, through super-enhancers (SEs), to implement the growth plate chondrocyte differentiation program [59]. *SOX9* mutations cause campomelic dysplasia (OMIM 114290), a skeletal dysplasia characterized by bowing of the long bones, small scapula, tracheobronchial narrowing, sex reversal, and kyphoscoliosis [60]. In adult mice, *SOX9* was reported to regulate *GPR126* gene expression and control the homeostasis of connective tissues such as the growth plate, articular cartilage, and the intervertebral disk [61]. Very long-range cis-regulatory elements controlling tissue-specific *SOX9* expression have been reported [62, 63]. The LD block containing rs12946942 has recently been defined as a susceptibility locus of prostate cancer [64]. The block contains six enhancer elements, one of which forms a long-range chromatin loop to *SOX9* in a prostate cancer cell line. Two SNPs within the enhancer shown to direct allele-specific gene expression, suggesting variants in the region, may similarly predispose to AIS by controlling scoliosis-related spatiotemporal SOX9 expression.

KCNJ2 encodes a potassium channel, a component of the inward rectifier current (IK1) [65]. *KCNJ2* mutations cause a cardiodysrhythmic type of periodic paralysis known as Andersen-Tawil syndrome (ATS; OMIM 170390) that is also characterized by ventricular arrhythmias, periodic paralysis, facial and skeletal dysmorphism including hypertelorism, small mandible, cleft palate, syndactyly, clinodactyly, and scoliosis [65, 66]. The 17q24.2–q24.3 microdeletion syndrome whose deletion that removes *KCNJ2* and rs12946942 exhibited skeletal malformations similar to ATS, including progressive scoliosis [67]. However, a similar microdeletion that includes *KCNJ2*, but not rs12946942, has no scoliosis phenotype [68]. Further studies are necessary to identify the causal gene in this locus.

To identify genes associated with AIS progression, Ogura et al. conducted a GWAS followed by a replication study using a total of 2543 AIS subjects who were evaluated for curve progression: the progression was defined as that with a Cobb angle ≥40° regardless of skeletal maturity and nonprogression as that with a Cobb angle ≤30° in skeletally mature patients [69]. These criteria are based on the natural history of AIS, wherein a curve with Cobb angle ≥40° may be progressive even after skeletal maturity, while that ≤30° is nonprogressive [70]. Curves between 31° and 39° were excluded to more clearly delineate the two groups, given that the Cobb method has 4–10° of intra- or interobserver error [71]. As a result, a novel locus on chromosome 11q14.1 was identified with genome-wide level of significant association ($P = 1.98 \times 10^{-9}$, odds ratio = 1.56). Subsequent in silico and in vitro analyses identified a functional variant, rs35333564 in *MIR4300HG*, the host gene of a microRNA, MIR4300. The genomic region containing rs35333564 had enhancer activity, which was decreased in its risk allele, suggesting that the decrease of MIR4300 is related to AIS progression.

The ability to objectively measure deformity progression (e.g., by Cobb angle measure) may provide the opportunity for genetic studies treating AIS as a quantitative trait as noted. Specifically, these analyses could consider Cobb angle at the time of surgical intervention or skeletal maturity as an endpoint measure. Survival analysis methods applied to longitudinal data (i.e., curve progression) might also be employed to identify factors that control progression rates [72]. In these methods, input variables may include SNP genotypes, gender, ethnicity, and initial measures of age, curve magnitude, curve pattern, bone age, etc., with the outcome variable being curve magnitude at a later time point. In this way, it may be possible to identify combinations of genotypes and phenotypes that classify progressive AIS. Both quantitative and qualitative studies should also benefit with improvements in imaging modalities to characterize the three dimensions of scoliosis in time.

Defining Rare Genetic Risk Factors for AIS

As with many complex disorders, genetic studies of AIS have been largely technology-driven. Consequently, while the majority of research has centered on common risk variants identified by GWAS, robust, massively parallel sequencing methods that query the entire exome or genome are increasingly cost-effective for discovering rare AIS risk alleles that are expected to contribute substantially to the disease. Rare disease alleles may be single nucleotide variants (SNVs) or small insertions/deletions (indels), and their discovery is bolstered by large-scale publicly available reference sets such as the genome aggregation database (gnomAD) that provide summary data for sequenced alleles [73] (Table 6.1). The search space for rare genetic risk factors is often limited to the exome by the so-called whole exome sequencing (WES) for practical reasons, i.e., interpretability. As the costs of sequencing decrease, the field is shifting more toward whole genome sequencing (WGS) and simultaneously improving the tools for identifying and interpreting noncoding rare disease variants that are rapidly coming online [74]. Association of rare variants with disease in a population is typically tested by statistically measuring the burden of rare alleles per gene (i.e., coding sequences) or genetic network as compared to controls [75]. This approach is expected to evolve in parallel with the functional annotation of the noncoding genome through efforts such as the Encyclopedia of DNA Elements (ENCODE) [76] and other investigations of regulatory elements in the human genome.

The contribution of rare variants in unrelated AIS populations was recently tested in cohorts of unrelated AIS cases and controls by WES and gene-based association [77]. Burden tests of WES in 91 AIS cases and 337 controls showed *FBN1* and *FBN2* encoding fibrillin-1 and fibrillin-2 as the most significantly associated genes with AIS, a result that was supported with further analyses in a larger cohort. It was noted that these mutations tended to be associated with tall stature, larger spinal curve, and slightly more Ghent criterion features [78], although the patients lack the cardiac manifestations or other complications that are typical of patients with Marfan syndrome (OMIM 154700) or congenital contractural arachnodactyly (OMIM 121050). Another study of 391 severe AIS cases and 843 controls of European ancestry examined the burden of rare variants by gene class or pathway and found rare mutation in extracellular matrix genes (ECM) in AIS ($P = 6 \times 10^{-9}$). In particular, novel coding variants in musculoskeletal collagen genes increased AIS risk by > twofold, with those in *COL11A2* being most strongly associated [79]. This suggests that rare, mildly deleterious variants in genes responsible for Mendelian connective tissue disorders may be enriched in individuals with idiopathic scoliosis.

Genomics and beyond

Although the majority of genetic risk in AIS awaits discovery, there is increasing evidence that variation in noncoding regulatory sequences contributes to disease susceptibility, presumably by altering spatiotemporal expression of genes such as *PAX1* and *LBX1*. Defining the regulatory landscape of these tissues, at multiple developmental time points, is therefore a major goal in AIS genetic research but challenging due to the paucity of information for AIS-relevant tissues such as disc, vertebral bone, paravertebral muscle, and spinal cord in existing resources such as ENCODE and GTEx [80]. Success will turn on the careful procurement and study of surgical specimens using methods such as RNA-seq and ChIP-seq or related methods such as ATAC-seq, Hi-C, capture Hi-C, ChIA-PET, Hi-ChIP, and proximity ligation-assisted ChIP-seq (PLAC-seq) to define regulatory elements and correlated expression at genome-wide scale [74]. Single-cell DNA and RNA sequencing is also a promising tool for expression analysis of heterogeneous samples [81]. Epigenetic changes, that is, alterations to the genome that do not involve changes in the DNA sequence itself, may also contribute to the genetic architecture of AIS. It is interesting to consider sex-specific methylation patterns in AIS, as DNA methylation patterns are known to vary between males and females. For example, Angelman/Prader-Willi syndrome and maternal uniparental disomy chromosome 14 can involve progressive scoliosis [82]. Integrating genomic, epigenetic, and clinical data is expected to not only provide a framework for understanding the mechanisms underlying this complex and heterogeneous disease but also for forging new diagnostic and perhaps prognostic clinical tools.

Assessing Risk and Potential for Personalized Medicine

Although the biologic contributions of individual genetic risk factors for AIS are expected to be small, studies of other complex disorders suggest that their aggregate effects are measurable and potentially informative. One approach to this calculates a "polygenic risk score (PRS)" that sums the effects of trait-associated common variants across the genome weighted by their effect sizes [14]. Theoretically a PRS might be used to identify high-risk individuals before disease onset or those individuals more likely to develop a particular disease subtype. For example, PRS have been shown to distinguish Crohn's disease from ulcerative colitis in inflammatory bowel disease [83]. One can envision that, as more AIS-associated variants are identified, they may be incorporated into a personal risk scoring system that may predict onset or perhaps aid in predicting risk of progression. Ultimately a primary goal is to understand how many genetic risk variants contribute biologically to increase AIS disease risk and to identify noninvasive interventions to reduce such risk. In this regard it is interesting to consider that the reduced penetrance and contributions from common risk variants suggest that there are inherent protections (i.e., "canalization") against AIS that may be harnessed [84]. As the "AIS genome" continues to be defined, it is also expected that genotype-phenotype correlations may emerge. An example is the association between mild deleterious *FBN1* and *FBN2* mutations and specific clinical features described above. Perhaps such AIS subgroups will become distinct clinical entities that may be described by a new nomenclature. Much more exciting is the prospect of personalized medicine, i.e., the possibility that AIS subgroups may be candidates for specific biologic therapies driven by individual genetic profiles.

AIS Consortium-Based Studies

The overall genetic architecture of AIS in terms of rare and common mutations and epigenetics will continue to emerge in parallel with ongoing and future discovery studies. The sum of the total genetic variation in AIS discovered to date probably explains less than 5% of disease susceptibility. Clearly, well-powered genetic studies are needed to define remaining contributions to this disease. Larger, more powerful GWAS are apparently warranted, given the promising association peaks that have been observed in exiting studies, but such studies are challenging to execute. Several tactics may help in this regard: [1] increasing sample size by recruiting larger cohorts, [2] utilizing more informative genotyping platforms, and [3] reducing genetic heterogeneity in the screened cohorts. Likewise, the continued discovery of rare disease variants will rest on experimental designs that increase power, for example, by applying family-based approaches or network-based analyses to large cohorts. In each scenario, we can expect to increase the power to detect associations by increasing the size and number of studied cohorts. Well-organized research consortia that can facilitate data harmonization and meta-analyses are one solution to this goal. The first such group, the International Consortium for Scoliosis Genetics (ICSG), was formed in 2012 for this purpose and subsequently produced the first AIS large-scale genetic meta-analysis, a study of the chromosome 10 *LBX1* locus in multiple cohorts [85]. In 2016 ICSG formally merged with the International Consortium for Vertebral Anomalies and Scoliosis (ICVAS) to become the

International Consortium for Spinal Genetics, Development, and Disease (ICSGDD) [86]. Ongoing ICSGDD efforts also will include genome-wide meta-analysis of existing datasets and organizational efforts to support the creation of much larger cohorts. Leading large consortium-driven studies of underrepresented ethnic/ancestral groups is a particularly high priority, as genetic studies of AIS thus far have centered on European/non-Hispanic white and East Asian populations that are unlikely to represent the full repertoire of genetic variation contributing to this global disease. Another key ICSGDD goal is to standardize and enhance AIS phenotyping that when correlated with specific genotypes may speed the identification of specific disease subgroups.

AIS Genetics: Summary and Future Research

Modern genomic tools are driving the discovery of genetic risk factors for AIS, primarily by GWAS. Rare variants are also expected to contribute to disease risk and are discoverable by well-powered sequence-based approaches. The total genetic risk profiles in individuals with AIS are expected to vary and may correlate with specific sub-phenotypes and/or disease progression. Integrated genomics hold the promise for identifying biologic networks that may be therapeutically targetable. The ability to scale genetic studies through emerging technologies and through consortium-sponsored collaboration will be key to defining the full genetic architecture of AIS.

Acknowledgments We thank the patients, families, and other individuals who have participated in AIS genetic research studies. We also thank Sarah Lassen from the Media Department at Texas Scottish Rite Hospital for Children for helping with figures.

References

1. Appleby J, Mitchell PD, Robinson C, Brough A, Rutty G, Harris RA, et al. The scoliosis of Richard III, last Plantagenet King of England: diagnosis and clinical significance. Lancet. 2014;383(9932):1944.
2. Herring J, editor. Tachdjian's Pediatric Orthopaedics. 5th ed. Philadelphia: WB Saunders; 2013.
3. Hresko MT. Clinical practice. Idiopathic scoliosis in adolescents. N Engl J Med. 2013;368(9):834–41.
4. Carter CO, Evans KA. Inheritance of congenital pyloric stenosis. J Med Genet. 1969;6(3):233–54.
5. Kruse LM, Buchan JG, Gurnett CA, Dobbs MB. Polygenic threshold model with sex dimorphism in adolescent idiopathic scoliosis: the Carter effect. J Bone Joint Surg Am. 2012;94(16):1485–91.
6. Antonarakis SE, Chakravarti A, Cohen JC, Hardy J. Mendelian disorders and multifactorial traits: the big divide or one for all? Nat Rev Genet. 2010;11(5):380–4.

7. Altshuler D, Daly MJ, Lander ES. Genetic mapping in human disease. Science. 2008;322(5903):881–8.
8. Spielman RS, Ewens WJ. The TDT and other family-based tests for linkage disequilibrium and association. Am J Hum Genet. 1996;59(5):983–9.
9. Mailman MD, Feolo M, Jin Y, Kimura M, Tryka K, Bagoutdinov R, et al. The NCBI dbGaP database of genotypes and phenotypes. Nat Genet. 2007;39(10):1181–6.
10. Tryka KA, Hao L, Sturcke A, Jin Y, Wang ZY, Ziyabari L, et al. NCBI's Database of Genotypes and Phenotypes: dbGaP. Nucleic Acids Res. 2014;42(Database issue):D975–9.
11. Freathy RM, Mook-Kanamori DO, Sovio U, Prokopenko I, Timpson NJ, Berry DJ, et al. Variants in ADCY5 and near CCNL1 are associated with fetal growth and birth weight. Nat Genet. 2010;42(5):430–5.
12. Hirata MNA, Kamatani Y, Ninomiya T, Tamakoshi A, Yamagata Z, Kubo M, Muto K, Kiyohara Y, Mushiroda T, Murakami Y, Yuji K, Furukawa Y, Zembutsu H, Tanaka T, Ohnishi Y, Nakamura Y, BioBank Japan Coperative Hospital Group, Matsuda K. Overview of BioBank Japan follow-up data in 32 diseases. J Epidemiol. 2017;27(3S):S22–S8.
13. Fuchsberger C, Flannick J, Teslovich TM, Mahajan A, Agarwala V, Gaulton KJ, et al. The genetic architecture of type 2 diabetes. Nature. 2016;536(7614):41–7.
14. Pasaniuc B, Price AL. Dissecting the genetics of complex traits using summary association statistics. Nat Rev Genet. 2017;18(2):117–27.
15. Hindorff LA, Junkins H.A, Mehta JP, Manolio TA. A Catalog of Published Genome-Wide Association Studies. Available at: http://www.genome.gov/gwastudies. Accessed [date of access].
16. Hindorff LA, Morales J (European Bioinformatics Institute), Junkins HA, Hall PN, Klemm AK, and Manolio TA. A Catalog of Published Genome-Wide Association Studies. Available from: http://www.genome.gov/gwastudies.
17. Takahashi Y, Kou I, Takahashi A, Johnson TA, Kono K, Kawakami N, et al. A genome-wide association study identifies common variants near LBX1 associated with adolescent idiopathic scoliosis. Nat Genet. 2011;43(12):1237–40.
18. Jiang H, Qiu X, Dai J, Yan H, Zhu Z, Qian B, et al. Association of rs11190870 near LBX1 with adolescent idiopathic scoliosis susceptibility in a Han Chinese population. Eur Spine J. 2013;22(2):282–6.
19. Gao W, Peng Y, Liang G, Liang A, Ye W, Zhang L, et al. Association between common variants near LBX1 and adolescent idiopathic scoliosis replicated in the Chinese Han Population. PLoS One. 2013;8(1):e53234.
20. Londono D, et al. A meta-analysis identifies adolescent idiopathic scoliosis association with LBX1 locus in multiple ethnic groups. J Med Genet. 2014;6:401–6.
21. Jagla K, Frasch M, Jagla T, Dretzen G, Bellard F, Bellard M. Ladybird, a new component of the cardiogenic pathway in Drosophila required for diversification of heart precursors. Development. 1997;124(18):3471–9.
22. Jagla K, Jagla T, Heitzler P, Dretzen G, Bellard F, Bellard M. Ladybird, a tandem of homeobox genes that maintain late wingless expression in terminal and dorsal epidermis of the Drosophila embryo. Development. 1997;124(1):91–100.
23. Jagla K, Dolle P, Mattei MG, Jagla T, Schuhbaur B, Dretzen G, et al. Mouse Lbx1 and human LBX1 define a novel mammalian homeobox gene family related to the Drosophila lady bird genes. Mech Dev. 1995;53(3):345–56.
24. Gross MK, Moran-Rivard L, Velasquez T, Nakatsu MN, Jagla K, Goulding M. Lbx1 is required for muscle precursor migration along a lateral pathway into the limb. Development. 2000;127(2):413–24.
25. Brohmann H, Jagla K, Birchmeier C. The role of Lbx1 in migration of muscle precursor cells. Development. 2000;127(2):437–45.
26. Schafer K, Neuhaus P, Kruse J, Braun T. The homeobox gene Lbx1 specifies a subpopulation of cardiac neural crest necessary for normal heart development. Circ Res. 2003;92(1):73–80.

27. Gross MK, Dottori M, Goulding M. Lbx1 specifies somatosensory association interneurons in the dorsal spinal cord. Neuron. 2002;34(4):535–49.
28. Muller T, Brohmann H, Pierani A, Heppenstall PA, Lewin GR, Jessell TM, et al. The homeodomain factor lbx1 distinguishes two major programs of neuronal differentiation in the dorsal spinal cord. Neuron. 2002;34(4):551–62.
29. Guo L, Yamashita H, Kou I, Takimoto A, Meguro-Horike M, Horike S, et al. Functional investigation of a non-coding variant associated with adolescent idiopathic scoliosis in zebrafish: elevated expression of the ladybird homeobox gene causes body axis deformation. PLoS Genet. 2016;12(1):e1005802.
30. Kou I, Takahashi Y, Johnson TA, Takahashi A, Guo L, Dai J, et al. Genetic variants in GPR126 are associated with adolescent idiopathic scoliosis. Nat Genet. 2013;45:676–9.
31. Xu JF, Yang GH, Pan XH, Zhang SJ, Zhao C, Qiu BS, et al. Association of GPR126 gene polymorphism with adolescent idiopathic scoliosis in Chinese populations. Genomics. 2015;105(2):101–7.
32. Soranzo N, Rivadeneira F, Chinappen-Horsley U, Malkina I, Richards JB, Hammond N, et al. Meta-analysis of genome-wide scans for human adult stature identifies novel Loci and associations with measures of skeletal frame size. PLoS Genet. 2009;5(4):e1000445.
33. Shukunami C, Shigeno C, Atsumi T, Ishizeki K, Suzuki F, Hiraki Y. Chondrogenic differentiation of clonal mouse embryonic cell line ATDC5 in vitro: differentiation-dependent gene expression of parathyroid hormone (PTH)/PTH-related peptide receptor. J Cell Biol. 1996;133(2):457–68.
34. Ikegawa S. Genomic study of adolescent idiopathic scoliosis in Japan. Scoliosis Spinal Disord. 2016;11:5.
35. Langenhan T, Aust G, Hamann J. Sticky signaling--adhesion class g protein-coupled receptors take the stage. Sci Signal. 2013;6(276):re3.
36. Monk KR, Naylor SG, Glenn TD, Mercurio S, Perlin JR, Dominguez C, et al. A G protein-coupled receptor is essential for Schwann cells to initiate myelination. Science. 2009;325(5946):1402–5.
37. Monk KR, Oshima K, Jors S, Heller S, Talbot WS. Gpr126 is essential for peripheral nerve development and myelination in mammals. Development. 2011;138(13):2673–80.
38. Waller-Evans H, Promel S, Langenhan T, Dixon J, Zahn D, Colledge WH, et al. The orphan adhesion-GPCR GPR126 is required for embryonic development in the mouse. PLoS One. 2010;5(11):e14047.
39. Geng FS, Abbas L, Baxendale S, Holdsworth CJ, Swanson AG, Slanchev K, et al. Semicircular canal morphogenesis in the zebrafish inner ear requires the function of gpr126 (lauscher), an adhesion class G protein-coupled receptor gene. Development. 2013;140(21):4362–74.
40. Karner CM, Long F, Solnica-Krezel L, Monk KR, Gray RS. Gpr126/Adgrg6 deletion in cartilage models idiopathic scoliosis and pectus excavatum in mice. Hum Mol Genet. 2015;24(15):4365–73.
41. Ogura Y, Kou I, Miura S, Takahashi A, Xu L, Takeda K, et al. A functional SNP in BNC2 is associated with adolescent idiopathic scoliosis. Am J Hum Genet. 2015;97(2):337–42.
42. Sharma S, Londono D, Eckalbar WL, Gao X, Zhang D, Mauldin K, et al. A PAX1 enhancer locus is associated with susceptibility to idiopathic scoliosis in females. Nat Commun. 2015;6:6452.
43. Wallin J, Wilting J, Koseki H, Fritsch R, Christ B, Balling R. The role of Pax-1 in axial skeleton development. Development. 1994;120(5):1109–21.
44. Zhu Z, Tang NL, Xu L, Qin X, Mao S, Song Y, et al. Genome-wide association study identifies new susceptibility loci for adolescent idiopathic scoliosis in Chinese girls. Nat Commun. 2015;6:8355.
45. Bharti S, Handrow-Metzmacher H, Zickenheiner S, Zeitvogel A, Baumann R, Starzinski-Powitz A. Novel membrane protein shrew-1 targets to cadherin-mediated junctions in polarized epithelial cells. Mol Biol Cell. 2004;15(1):397–406.

46. Schubert FR, Tremblay P, Mansouri A, Faisst AM, Kammandel B, Lumsden A, et al. Early mesodermal phenotypes in splotch suggest a role for Pax3 in the formation of epithelial somites. Dev Dyn. 2001;222(3):506–21.
47. Buckingham M, Relaix F. The role of Pax genes in the development of tissues and organs: Pax3 and Pax7 regulate muscle progenitor cell functions. Annu Rev Cell Dev Biol. 2007;23:645–73.
48. Young AP, Wagers AJ. Pax3 induces differentiation of juvenile skeletal muscle stem cells without transcriptional upregulation of canonical myogenic regulatory factors. J Cell Sci. 2010;123(Pt 15):2632–9.
49. Relaix F, Rocancourt D, Mansouri A, Buckingham M. A Pax3/Pax7-dependent population of skeletal muscle progenitor cells. Nature. 2005;435(7044):948–53.
50. Kullander K, Butt SJ, Lebret JM, Lundfald L, Restrepo CE, Rydstrom A, et al. Role of EphA4 and EphrinB3 in local neuronal circuits that control walking. Science. 2003;299(5614):1889–92.
51. Farlie PG, Dringen R, Rees SM, Kannourakis G, Bernard O. bcl-2 transgene expression can protect neurons against developmental and induced cell death. Proc Natl Acad Sci U S A. 1995;92(10):4397–401.
52. Moriishi T, Maruyama Z, Fukuyama R, Ito M, Miyazaki T, Kitaura H, et al. Overexpression of Bcl2 in osteoblasts inhibits osteoblast differentiation and induces osteocyte apoptosis. PLoS One. 2011;6(11):e27487.
53. Wu J, Qiu Y, Zhang L, Sun Q, Qiu X, He Y. Association of estrogen receptor gene polymorphisms with susceptibility to adolescent idiopathic scoliosis. Spine (Phila Pa 1976). 2006;31(10):1131–6.
54. Zhang HQ, Lu SJ, Tang MX, Chen LQ, Liu SH, Guo CF, et al. Association of estrogen receptor beta gene polymorphisms with susceptibility to adolescent idiopathic scoliosis. Spine (Phila Pa 1976). 2009;34(8):760–4.
55. Chen Z, Tang NL, Cao X, Qiao D, Yi L, Cheng JC, et al. Promoter polymorphism of matrilin-1 gene predisposes to adolescent idiopathic scoliosis in a Chinese population. Eur J Hum Genet. 2009;17(4):525–32.
56. Yeung HY, Tang NL, Lee KM, Ng BK, Hung VW, Kwok R, et al. Genetic association study of insulin-like growth factor-I (IGF-I) gene with curve severity and osteopenia in adolescent idiopathic scoliosis. Stud Health Technol Inform. 2006;123:18–24.
57. Miyake A, Kou I, Takahashi Y, Johnson TA, Ogura Y, Dai J, et al. Identification of a susceptibility locus for severe adolescent idiopathic scoliosis on chromosome 17q24.3. PLoS One. 2013;8(9):e72802.
58. Dy P, Wang W, Bhattaram P, Wang Q, Wang L, Ballock RT, et al. Sox9 directs hypertrophic maturation and blocks osteoblast differentiation of growth plate chondrocytes. Dev Cell. 2012;22(3):597–609.
59. Liu CF, Lefebvre V. The transcription factors SOX9 and SOX5/SOX6 cooperate genome-wide through super-enhancers to drive chondrogenesis. Nucleic Acids Res. 2015;43(17):8183–203.
60. Lekovic GP, Rekate HL, Dickman CA, Pearson M. Congenital cervical instability in a patient with camptomelic dysplasia. Child's Nerv Syst. 2006;22(9):1212–4.
61. Henry SP, Liang S, Akdemir KC, de Crombrugghe B. The postnatal role of Sox9 in cartilage. J Bone Miner Res. 2012;27(12):2511–25.
62. Wunderle VM, Critcher R, Hastie N, Goodfellow PN, Schedl A. Deletion of long-range regulatory elements upstream of SOX9 causes campomelic dysplasia. Proc Natl Acad Sci U S A. 1998;95(18):10649–54.
63. Gordon CT, Tan TY, Benko S, Fitzpatrick D, Lyonnet S, Farlie PG. Long-range regulation at the SOX9 locus in development and disease. J Med Genet. 2009;46(10):649–56.
64. Zhang X, Cowper-Sal lari R, Bailey SD, Moore JH, Lupien M. Integrative functional genomics identifies an enhancer looping to the SOX9 gene disrupted by the 17q24.3 prostate cancer risk locus. Genome Res. 2012;22(8):1437–46.
65. Tristani-Firouzi M, Etheridge SP. Kir 2.1 channelopathies: the Andersen-Tawil syndrome. Pflugers Arch. 2010;460(2):289–94.

66. Plaster NM, Tawil R, Tristani-Firouzi M, Canun S, Bendahhou S, Tsunoda A, et al. Mutations in Kir2.1 cause the developmental and episodic electrical phenotypes of Andersen's syndrome. Cell. 2001;105(4):511–9.
67. Lestner JM, Ellis R, Canham N. Delineating the 17q24.2-q24.3 microdeletion syndrome phenotype. Eur J Med Genet. 2012;55(12):700–4.
68. Blyth M, Huang S, Maloney V, Crolla JA, Karen Temple I. A 2.3Mb deletion of 17q24.2-q24.3 associated with 'Carney complex plus'. Eur J Med Genet. 2008;51(6):672–8.
69. KI OY, Takahashi Y, Takeda K, Minami S, Kawakami N, Uno K, Ito M, Yonezawa I, Kaito T, Yanagida H, Watanabe K, Taneichi H, Harimaya K, Taniguchi Y, Kotani T, Tsuji T, Suzuki T, Sudo H, Fujita N, Yagi M, Chiba K, Kubo M, Kamatani Y, Nakamura M, Matsumoto M, Japan Scoliosis Clinical Research Group, Watanabe K, Ikegawa S, Japan Scoliosis Clinical Research Group. A functional variant in MIR4300HG, the host gene of microRNA MIR4300 is associated with progression of adolescent idiopathic scoliosis. Hum Mol Genet. 2017;26(20):4086–92.
70. Weinstein SL. Natural history. Spine (Phila Pa 1976). 1999;24(24):2592–600.
71. Carman DL, Browne RH, Birch JG. Measurement of scoliosis and kyphosis radiographs. Intraobserver and interobserver variation. J Bone Joint Surg. 1990;72(3):328–33.
72. Londono D, Chen KM, Musolf A, Wang R, Shen T, Brandon J, et al. A novel method for analyzing genetic association with longitudinal phenotypes. Stat Appl Genet Mol Biol. 2013;12(2):241–61.
73. Lek M, Karczewski KJ, Minikel EV, Samocha KE, Banks E, Fennell T, et al. Analysis of protein-coding genetic variation in 60,706 humans. Nature. 2016;536(7616):285–91.
74. Chatterjee S, Ahituv N. Gene regulatory elements, major drivers of human disease. Annu Rev Genomics Hum Genet. 2017;18:45–63.
75. Lee S, Emond MJ, Bamshad MJ, Barnes KC, Rieder MJ, Nickerson DA, et al. Optimal unified approach for rare-variant association testing with application to small-sample case-control whole-exome sequencing studies. Am J Hum Genet. 2012;91(2):224–37.
76. Consortium EP, Bernstein BE, Birney E, Dunham I, Green ED, Gunter C, et al. An integrated encyclopedia of DNA elements in the human genome. Nature. 2012;489(7414):57–74.
77. Buchan JG, Alvarado DM, Haller GE, Cruchaga C, Harms MB, Zhang T, et al. Rare variants in FBN1 and FBN2 are associated with severe adolescent idiopathic scoliosis. Hum Mol Genet. 2014;23(19):5271–82.
78. Loeys BL, Dietz HC, Braverman AC, Callewaert BL, De Backer J, Devereux RB, et al. The revised Ghent nosology for the Marfan syndrome. J Med Genet. 2010;47(7):476–85.
79. Haller G, Alvarado D, McCall K, Yang P, Cruchaga C, Harms M, et al. A polygenic burden of rare variants across extracellular matrix genes among individuals with adolescent idiopathic scoliosis. Hum Mol Genet. 2016;25(1):202–9.
80. Project e. Enhancing GTEx by bridging the gaps between genotype, gene expression, and disease. Nat Genet. 2017;49:1664–70.
81. Grun D, van Oudenaarden A. Design and analysis of single-cell sequencing experiments. Cell. 2015;163(4):799–810.
82. Temple IK, Cockwell A, Hassold T, Pettay D, Jacobs P. Maternal uniparental disomy for chromosome 14. J Med Genet. 1991;28(8):511–4.
83. Cleynen I, Boucher G, Jostins L, Schumm LP, Zeissig S, Ahmad T, et al. Inherited determinants of Crohn's disease and ulcerative colitis phenotypes: a genetic association study. Lancet. 2016;387(10014):156–67.
84. Felix MA, Barkoulas M. Pervasive robustness in biological systems. Nat Rev Genet. 2015;16(8):483–96.
85. Londono D, Kou I, Johnson TA, Sharma S, Ogura Y, Tsunoda T, et al. A meta-analysis identifies adolescent idiopathic scoliosis association with LBX1 locus in multiple ethnic groups. J Med Genet. 2014;51(6):401–6.
86. Giampietro PF, Pourquié O, Raggio C, Ikegawa S, Turnpenny PD, Gray R, et al. Summary of the first inaugural joint meeting of the International Consortium for scoliosis genetics and the International Consortium for Vertebral Anomalies and Scoliosis, March 16–18, 2017, Dallas, Texas. Am J Med Genet A. 2017;176(1):253–6.

Chapter 7
Genetics and Functional Pathology of Idiopathic Scoliosis

Elizabeth A. Terhune, Erin E. Baschal, and Nancy Hadley Miller

Introduction

Idiopathic scoliosis (IS) is a structural lateral curve of the spine that affects approximately 2–3% of pediatric populations, with girls more severely affected than boys. Treatment options are currently limited to bracing, physical therapy, and spinal fusion surgery for severe progressive curves. The variability in clinical presentation, limited treatment options, and inability to detect those at risk for curve progression have confounded physicians as well as IS patients and their families. IS has long been recognized to have a familial or genetic component; however, the mechanisms underlying this heritability are largely unknown. Multiple studies to date have identified genetic variants that are associated with IS in specific cohorts. However, most of these associations, with the exception of variants in or near *LBX1* and *GPR126*, have not been able to be reproduced. The varied results of these studies are indications of the extreme genetic and phenotypic heterogeneity of this disorder. New technologies, including next-generation sequencing and improved animal models, hold promise for the discovery of additional mechanisms that cause IS. Identifying the genetic factors underlying IS may aid in the development of diagnostic screening tools and more effective treatment options for affected children.

E. A. Terhune · E. E. Baschal
Department of Orthopedics, University of Colorado Anschutz Medical Campus, Aurora, CO, USA

N. H. Miller (✉)
Department of Orthopedics, University of Colorado Anschutz Medical Campus, Aurora, CO, USA

Musculoskeletal Research Center, Children's Hospital Colorado, Aurora, CO, USA
e-mail: nancy.hadley-miller@ucdenver.edu

K. Kusumi, S. L. Dunwoodie (eds.), *The Genetics and Development of Scoliosis*, https://doi.org/10.1007/978-3-319-90149-7_7

A Genetic Basis for Idiopathic Scoliosis (IS)

A Familial Basis of IS

The hereditary basis of IS was established as early as the 1930s, when scoliosis was identified in five generations in one family [1]. Decades later, clinical observations and population studies documented a higher prevalence of scoliosis among relatives of affected individuals compared to the general population [1–7]. Specifically, Wynne-Davies observed that relatives of individuals with IS were at a higher risk for developing the disease, reporting that IS was present in 11% of first-degree, 2.4% of second-degree, and 1.4% of third-degree relatives [2, 8]. In another study, Bonaiti et al. observed that approximately 40% of IS cases were familial across multiple populations [9]. More recently, in an analysis of a unique database of a Mormon population in Utah (GenDB), 97% of IS patients were determined to be of familial origin [10]. Many researchers within the IS research community have since hypothesized that *IS is likely due to multiple inherited risk alleles in tandem with environmental risk factors.*

Twin Concordance Rates

Studies of monozygotic and dizygotic twins have provided further evidence supporting a genetic basis of IS. Concordance is defined as both twins having the disease, and higher concordance rate in monozygotic twins compared to dizygotic twins is an indication that a disease has a genetic component. The concordance for IS is approximately 73% for monozygotic twins and 36% for dizygotic twins [7, 11–17], indicating there is a strong genetic component to the disease. At first, this may appear confusing and contradictory, as dizygotic twins appear to have a threefold higher concordance rate than that reported by Riseborough and Wynne-Davies for first-degree relatives [8]. However, upon further examination, the higher rate in dizygotic twins may be due to the differences in rates of radiographic confirmation of the scoliotic curve. Radiographic confirmation of scoliosis may be higher among twins, as the diagnosis of one twin may lead to inquiries into the curvature status of the other twin. This radiographic confirmation may be less likely within first-degree relatives, and thus first-degree relatives may have an artificially low concordance rate. There could also be an in utero environmental component to IS that has not yet been identified. Although these factors may confound the interpretation of the twin data, the high concordance rates in both monozygotic and dizygotic twins are an indication of a strong genetic component to IS.

Multiple Modes of Inheritance

To date, multiple modes of genetic inheritance have been reported in investigations of IS families, including X-linked [18, 19], multifactorial [2, 20, 21], and autosomal dominant [22–25]. Rather than presenting conflicting information, these reports serve as examples of the *diverse spectrum of inheritance models for IS* that can exist between families. Figure 7.1 presents two examples of commonly seen IS

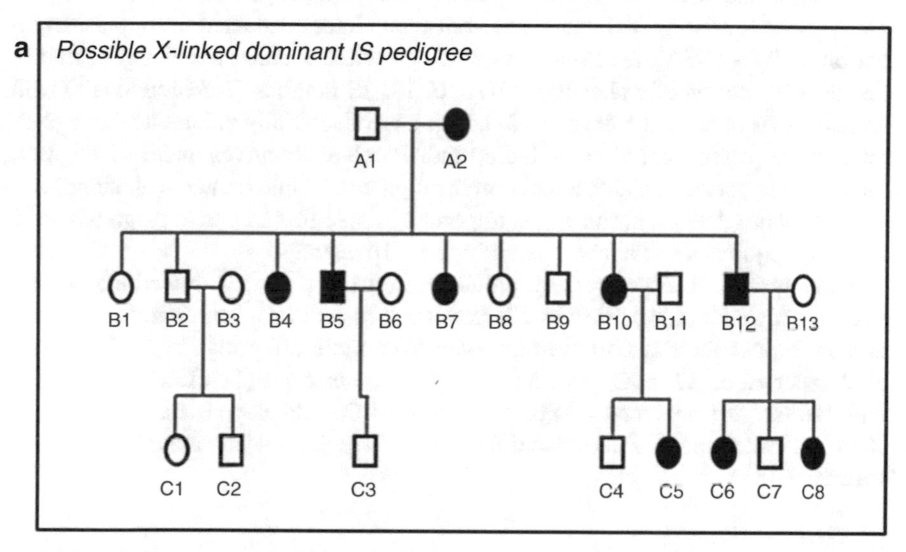

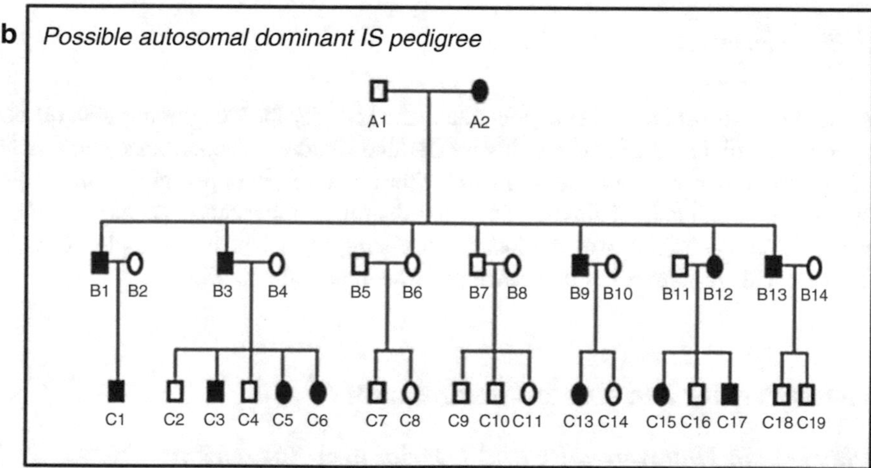

Fig. 7.1 Examples of distinct but typical IS pedigrees. (**a**) A likely X-linked dominant pedigree of a multigenerational IS family. Note the lack of male-to-male transmission. (**b**) IS pedigree displaying male-to-male transmission and a possible autosomal dominant inheritance pattern

pedigrees— one exhibiting an apparent X-linked dominant mode of inheritance and the other showing an apparent autosomal dominant mode of inheritance.

Wynne-Davies suggested that IS was likely inherited in either a dominant or multifactorial mode of inheritance, based on their early studies in the 1960s–1970s with over 2000 individuals [2, 8]. Aksenovich [26] also suggested an autosomal dominant inheritance pattern for IS with sex-dependent penetrance, after conducting an analysis of families with probands with moderate to severe scoliosis (>25 degrees). However, when the authors analyzed probands with mild scoliosis (<25 degrees), multiple damaging alleles (genetic heterogeneity) were attributed as the likely cause for the IS. This same research group later conducted a complex segregation analysis (CSA), a technique used to determine whether a major gene underlies the distribution of a phenotypic trait, of 101 IS families (703 individuals) with spinal curves of at least 5 degrees [27]. This analysis initially did not lead to a clear inheritance pattern, but after excluding individuals with curves under 11 degrees, the results supported an autosomal dominant pattern of inheritance with significant rates of incomplete penetrance. It is important to note that a clinical diagnosis of IS typically requires a spinal curvature of at least 10 degrees.

Other groups observed pedigrees that appear to display an X-linked inheritance pattern. This pattern is defined by a lack of male-to-male transmission, as males are unable to pass their X chromosome onto their male offspring. In 1972, Cowell et al. reported on 17 families with this mode of inheritance [18]. Later, Miller [19] and Justice [28] analyzed a subset of families that displayed characteristics of X-linked inheritance and found that the Xq23–26 region may be linked to IS in this subset.

Section Summary

A genetic basis for IS has been well-established for decades, beginning with observational familial studies in the 1930s by Garland. Studies of inheritance patterns of IS have proven to vary both between and within families, with pedigrees supporting autosomal dominant, X-linked, and multifactorial inheritance patterns. Taken together, data acquired over the last century suggest that IS is a complex genetic disorder with variations in inheritance patterns in affected families.

Heterogeneity and Confounding Factors

Phenotypic Heterogeneity and Overlapping Conditions

Significant phenotypic and genetic heterogeneity across patient cohorts have caused significant challenges for determining the genetic factors underlying IS. Non-idiopathic scoliosis is part of the disease phenotype for several musculoskeletal conditions, including Marfan syndrome and osteogenesis imperfecta, and early studies of IS may not have fully excluded these individuals from their cohorts [1, 3, 29, 30]. Other studies did not detail specific criteria for affected versus unaffected scoliosis status or were unable to consistently obtain spinal radiographs from study participants [2, 7, 8, 20, 22, 31]. Well-documented diagnostic criteria and radiographic confirmation of curve magnitude can aid in the distinction between true affected and unaffected individuals, and radiographic confirmation can potentially reduce the rate of false-negative diagnoses made from clinical observations alone. In addition, thorough patient records are necessary for phenotypic subtype analyses within large study cohorts, and desired clinical information may include age of onset, curve severity, curve progression, family history of IS, and lifestyle or environmental factors. Detailed phenotypic characterization, including radiographs, is important for assigning both affected and unaffected status and will allow for a better understanding of IS.

Genetic Heterogeneity

Likewise, genetic heterogeneity of patient populations is a significant challenge for IS researchers to overcome. Genetic heterogeneity is defined as a phenotype that arises by different genetic mechanisms. In the context of IS, this means that mutations in multiple genes could each give rise to a similar phenotype— a lateral spinal curvature that is currently labeled as IS. However, once the underlying genetic causes of IS are determined, it is possible that "idiopathic scoliosis" will be broken down into different subtypes based on phenotypic and genetic differences. In practice, it can be difficult to identify a clear inheritance pattern in families with high degrees of genetic heterogeneity, and these families may also appear to have high rates of incomplete penetrance. A major concern with genetic heterogeneity in IS is that individuals are treated as having the exact same condition for experimental and analysis purposes. This could confound both linkage and association studies. A full understanding of the genetic spectrum of IS may also be important for treatment options for the different disease subsets.

Section Summary

IS is marked by a high degree of genetic and phenotypic heterogeneity, creating a significant challenge for researchers trying to unravel the genetic regions and mechanisms underlying this condition. Strict diagnostic criteria, particularly radiographic confirmation of IS status, are required to distinguish true IS from other conditions, as well as to identify true negative controls.

Overview of IS Genetic Findings

Candidate Gene Studies

Prior to the availability of next-generation sequencing technologies, candidate gene approaches were used in order to analyze protein-coding genes thought to be important to the physiological basis of IS. As scoliosis is a phenotypic component of many connective tissue disorders, several extracellular matrix (ECM) genes including collagens, elastin, fibrillin, and aggrecan were studied in IS individuals and families [25, 32]. However, these studies largely resulted in negative findings, as identified variants in these genes often did not segregate with the disease phenotype. Later studies using more modern sequencing technologies, however, did reveal certain significant associations between variants in ECM genes and the IS phenotype (see *Next-Generation Sequencing*).

Linkage Analysis

After advances in genetic technologies in the early 1990s, researchers were able to screen the entire genome for known genetic markers or polymorphisms evenly spaced throughout the chromosome. One such analysis resulting from this advancement was *linkage analysis,* which relies on the concept of genetic *linkage*, or the tendency of certain genes or genetic regions to be inherited together due to physical proximity on a chromosome. Linkage studies typically analyze certain genetic markers in large families with a disease and seek to identify marker alleles that are only present in affected individuals. From there, the candidate region can be narrowed down further with additional fine mapping within the family members, including within candidate genes that may be present in the region, often using single nucleotide polymorphisms (SNPs). In parametric linkage analysis, LOD (*l*ogarithm of the *od*ds) scores are reported to assess whether the allele segregating with the disease phenotype is due to linkage or due to chance. Parametric linkage requires specification of allele frequencies, penetrance, and an inheritance pattern. Nonparametric linkage analysis does not make any assumptions about the disease model and reports the probability of family members sharing alleles identical by descent. It should be noted that the presence of linkage or association of a locus with a disease does not prove causality. Results need to be validated within independent cohorts and then studied for functional importance.

In 2000, Wise et al. conducted nonparametric linkage analysis on a multiplex IS family, which suggested four regions, on chromosomes 3, 6, 12, and 18, as potentially important for IS. After further study of regions 6, 10, and 18 in a second family, region 18q was determined to be the most important region for linkage, with a secondary area on chromosome 6p. Additionally, both families supported a common candidate on distal chromosome 10q [33]. In 2002, Chan et al. analyzed the three regions identified by Wise et al. in one multiplex family but were not able to replicate the linkage. The Chan group conducted a second genome-wide scan of seven families, which identified two regions of interest, a primary area on chromosome 19p13.3 and a secondary area on chromosome 2q [34]. Salehi et al. also conducted linkage analysis on one large multiplex family, which yielded a candidate region on chromosome 17p11 [35]. This region was of particular interest, as it contained several ECM genes.

In 2005, Miller et al. reported a large genetic linkage screen of 202 families (1198 individuals), and stratified families based on phenotypic subtypes and the apparent mode of inheritance, to decrease the heterogeneity within the population. Linkage analysis of families with an apparent autosomal dominant inheritance pattern yielded primary regions on chromosomes 6p, 6q, 9, 16, and 17, as well as secondary regions on chromosomes 1, 3, 5, 7, 8, 11, and 12. Similarly, in families with an apparent X-linked inheritance pattern, the Xq23–26 candidate region was identified. Stratification of the samples into families with an individual with *kyphoscoliosis* yielded significant regions on chromosomes 5 and 13, and analysis of families with an individual with a severe curve (>40 degrees) yielded a region on chromosome 19 [36].

Gao et al. produced evidence of linkage and association of the 8q12 region. Fine mapping association studies of this region revealed evidence of IS-associated haplotypes centered over exons 2–4 of the *CHD7* gene [37]. Interestingly, mutations in *CHD7* are associated with CHARGE syndrome, for which scoliosis is frequently part of the disease phenotype. However, an independent association study of 22 single nucleotide polymorphism (SNPs) in the *CHD7* gene in 244 IS families failed to replicate the *CHD7* finding [38].

Edery et al. performed a genome-wide scan of three large multigenerational IS families compatible with an autosomal dominant inheritance pattern [39]. The group was not able to replicate the previous findings for 19p13.3, 17p11.2, 9q34, and 18q in any of the three families. However, they observed disease co-segregation in the 3q12.1 and 5q13.3 loci in one family. Subsequent exome sequencing in this family narrowed the disease gene to *POC5*, and injection of patient-specific *POC5* mRNA into zebrafish embryos led to the development of an IS-like phenotype [40] (see *Functional Studies and Animal Models*).

Genome-Wide Association Studies

Genome-wide association studies (GWAS) use genetic data from large cohorts to test the association of a phenotype with a genotype, typically a SNP. The first GWAS for IS was reported in 2011 by Sharma et al. [41], which assayed 419 adolescent IS families with 327,000 SNPs. The authors found the strongest evidence for association with chromosome 3p26.3 SNPs in the proximity of the *CHL1* gene, which encodes an axon guidance protein related to ROBO3. Later that year, a GWAS of 1376 Japanese females with adolescent IS and 11,297 controls revealed a significant association with variants near the gene *LBX1* [42]. *LBX1* encodes the transcription factor ladybird homeobox 1, which is an important determinant of spinal cord neuron migration and cell fate choice [42]. Significant associations of SNPs near *LBX1* have since been reported in several additional studies, including those from Chinese, European, and French-Canadian cohorts [43–52]. The same research group who performed the Takahashi et al. GWAS expanded their original study cohort to 1819 cases and 25,939 controls of Japanese ancestry, which revealed a susceptibility locus within the G-protein-coupled receptor gene *GPR126* [53]. This association was replicated in two independent studies within IS cohorts of Chinese ancestry [54, 55]. Other IS GWAS have reported associations within *BCN2* [56], between *SOX9* and *KCNJ2* [57], near *PAX1* [58], and with several loci in or near genes involved in Wnt signaling [59]. Table 7.1 provides a summary of IS GWAS, linkage, and other association studies to date.

High-Throughput Sequencing

The advent and adoption of high-throughput ("next-generation") sequencing technologies in the 2000s spurred a genetic revolution by allowing researchers to sequence whole genomes or exomes in a fraction of the time and cost of traditional sequencing methods. Exome sequencing captures the 1–2% of the human genome that is predicted to be protein-coding and allows for the identification of both rare and common variants in these protein-coding genes. Exome sequencing is based on the hypothesis that variants in protein-coding regions are more likely to have functional effects that could cause the disease and has been used successfully to identify causative variants for many diseases, particularly those that are monogenic.

Exome sequencing has been used in multiple IS studies to identify candidate variants and genes. In 2014, Buchan et al. reported an exome sequencing study of 91 unrelated individuals of European ancestry with severe scoliosis (>40 degrees), which revealed the variant burden in *FBN1* as most associated with adolescent IS. Subsequent sequencing of both *FBN1* and *FBN2* in a larger cohort showed a significant enrichment of rare variants in both genes within Caucasian individuals with severe scoliosis (7.6%) compared with in-house controls (2.4%) ($p = 5.46 \times 10^{-4}$) and exome sequencing project controls (2.3%) ($p = 1.48 \times 10^{-6}$) [60]. These findings were also replicated in an independent Han Chinese cohort ($p = 0.0376$), suggesting that these rare variants might be useful markers of curve progression. In 2015, Baschal et al. reported that rare variants in *HSPG2*, which encodes the ECM protein perlecan, were associated with the IS phenotype in a multigenerational AIS family. One particular rare variant, p.Asn786Ser, was also overrepresented in an additional cohort of 100 unrelated IS cases as compared to controls ($p = 0.024$) [61]. ECM variants were further implicated by Haller et al. in an exome sequencing study of 391 severe AIS cases and 843 controls of European ancestry [62]. Novel non-synonymous/splice-site variants in ECM genes were significantly enriched in cases versus controls ($p = 6 \times 10^{-9}$); furthermore, novel variants in

Table 7.1 Genetic studies of idiopathic scoliosis (IS) cohorts using linkage or association analyses

Study	Approach	Number and type of sample	Region	Candidate gene(s)/ marker(s)	Significance (*p* value)
Carr et al. 1992 [23]	Candidate gene, linkage	4 families	17q21	*COL1A1*	NS
			7q22	*COL1A2, COL2A1*	NS
Miller et al. 1996 [25]	Candidate gene, linkage	11 families	15q21.1	*FBN1*	NS
			7q11	*ELN*	NS
			7q22	*COL1A2*	NS
Wise et al. 2000 [33]	Linkage	2 families	6q	NA	0.023, NS
			Distal 10q	NA	0.0193, 0.033
			18q	NA	0.0023, NS
Morcuende et al. 2003 [75]	Candidate gene, linkage	47 families	4q	*MTNR1A*	NS
Inoue et al. 2002 [88]	Candidate gene	304 cases	6q25	*ESR1*	0.002
Chan et al. 2002 [34]	Linkage	7 families	19p13.3	D19S894–D19S1034	0.00001 (4.48)[4]
			2q13–2q22.3	D2S160–D2S151	0.0049 (1.72)[4]
Salehi et al. 2002 [35]	Linkage	1 family	17p11–17q11.2	D17S799–D17S925	0.0001 (3.2)[4]
Justice et al. 2003 [28]	Linkage	51 families	Xq23–Xq26.1	DXS6804–DXS1047	0.0014 (2.23)[4]
Miller et al. 2005 [36]	Linkage	202 families	6p	F13A1–D6S2439	0.01215
			6q16	D6S1031–D6S1021	0.00215
			9q32–9q34	D9S938–D9S1838	0.00055
			16q11–16q12	D16S764–D16S3253	0.00025
			17p11–17q11	D17S1303–D17S1293	0.0025
Alden et al. 2006 [89]	Linkage	72 families	19p13	D19S591–D19S1034	0.013565
Miller et al. 2006 [89]	Linkage, association	7 families	5q13	D5S417–D5S807	0.00173
			13q13.3	D13S305–D13S788	0.00013
			13q32	D13S800–D13S779	0.00013
Yeung et al. 2006 [90]	Candidate gene	506 cases	12q22	*IGF1*	NS

(continued)

Table 7.1 (continued)

Study	Approach	Number and type of sample	Region	Candidate gene(s)/ marker(s)	Significance (*p* value)
Wu et al. 2006 [91]	Candidate gene	202 cases	6q25	*ESR1*	0.001
Tang et al. 2006 [92]	Candidate gene	540 cases, 260 controls	6q25	*ESR1*	NS
Qiu et al. 2006 [93]	Candidate gene	473 AIS, 311 controls	11q21	*MTNR1B*	NS
Montanaro et al. 2006 [94]	Candidate gene, linkage	81 trios	1p35	*MATN1*	0.024
Marosy et al. 2006 [32]	Candidate gene, linkage	58 families	15q25–26	*AGC1*	NS
Gao et al. 2007 [37]	Linkage, association	52 families	8q12	*CHD7*	0.005
Ocaka et al. 2008 [95]	Linkage	25 families	9q31.2–q34.2	D9S930–D9S1818	0.00004
			17q25.3-qtel	D17S1806	0.00001
Raggio et al. 2009 [96]	Linkage, association	7 families	12p	D12S1608–D12S1674	Unknown
Marosy et al. 2010 [97]	Linkage, association	3 families (triple curves)	6p	D6S1043-D6S474	<0.001
			10q	D10S2325–D10S1423 and D10S1765–D10S1239	<0.001
Clough et al. 2010 [98]	Linkage, association	17 families (males)	17p	D17S975, D17S2196	<0.05
Edery et al. 2011 [39]	Linkage, association	3 families (1 family with disease co-segregation)	3q12.1	D3S3690–D3S3045,	<0.001
			5q13.3	D5S2851–D5S1397	<0.001
Sharma et al. 2011 [41]	GWAS	419 families	3p26.3	*CHL1*	2.58×10^{-8}
Takahashi et al. 2011 [42]	GWAS	1376 AIS and 11,297 controls	10q24.31	*LBX1*	1.24×10^{-19}
Gao et al. 2013 [44]	Association, replication	513 AIS and 440 controls	10q24.31	*LBX1*	5.09×10^{-5}–1.17×10^{-8}
Kou et al. 2013 [53]	GWAS	1819 AIS and 25,939 controls	6q24.1	*GPR126*	2.25×10^{-10}
Miyake et al. 2013 [57]	GWAS	554 AIS (severe) and 1474 controls	17q24.3	*SOX9, KCNJ2*	4.00×10^{-18}

(continued)

Table 7.1 (continued)

Study	Approach	Number and type of sample	Region	Candidate gene(s)/marker(s)	Significance (p value)
Londono et al. 2014 [46]	Meta-analysis, replication	9 cohorts	10q24.31	*LBX1*	1.22×10^{-43} (*for rs11190870*)
Zhu et al. 2015 [48]	GWAS	4317 AIS and 6016 controls	1p36.32	*AJAP1*	2.95×10^{-9}
			2q36.1	*PAX3, EPHA4*	7.59×10^{-13}
			18q21.33	*BCL2*	2.22×10^{-12}
Ogura et al. 2015 [56]	GWAS	2109 AIS and 11,140 controls	9p22.2	*BNC2*	2.46×10^{-13}
Sharma et al. 2015 [58]	GWAS	3102 individuals	20p11.22	*PAX1*	6.89×10^{-9}
Zhu et al. 2017 [59]	GWAS	5953 AIS and 8137 controls	2p14	*MEIS1*	1.19×10^{-13}

P-value denotes most significant published value.
NS not significant.

musculoskeletal collagen genes were present in 32% of AIS cases versus 17% of controls. Patten et al. combined genetic linkage data with exome sequencing of one large IS family to identify a rare variant in *POC5*, a centrosomal protein, as associated with the phenotype [40]. In 2016, Li et al. performed exome sequencing on a large family with IS to identify *AKAP2*, a gene encoding a cAMP regulatory protein that associates with the actin cytoskeleton [63]. In a recent study, Gao et al. combined linkage data from a three-generation IS family of Chinese decent with exome sequence data in a discovery cohort of 20 AIS individuals and 86 controls, which showed a significant association of the IS phenotype with three missense variants in the *MAPK7* gene [64]. *MAPK7* encodes a nuclear transport protein, and in vitro experiments demonstrated that the three missense variants each disrupted nuclear translocation in cellular models. Table 7.2 provides a non-exhaustive summary of IS studies using next-generation sequencing technologies.

Transcriptomics and Other Approaches

Although an individual's genomic DNA is generally identical across tissues, their messenger RNA (mRNA) will vary from tissue to tissue to create unique expression signatures, collectively defined as the *transcriptome*. Several groups have compared the gene expression in relevant cell types between IS and control individuals, with the aim of identifying biological differences that may more accurately reflect what is occurring at the protein level of the cell. Osteoblasts, bone cells that form mineralized matrix, have been analyzed by several IS research groups due to their importance to skeletal growth and maintenance, as well as their ability to be collected

Table 7.2 Genetic studies of idiopathic scoliosis (IS) cohorts using next-generation sequencing (NGS)

Study	Approach	Number and type of samples	Candidate gene(s)	Significance (*p* value)
Buchan et al. 2014b [60]	Exome	852 AIS and 669 controls	*FBN1, FBN2*	5.46×10^{-4}
Baschal et al. 2015 [61]	Exome	1 family with validation in 240 AIS/4679 controls	*HSPG2*	0.024
Patten et al. 2015 [40]	Exome	1 family with validation in 40 families	*POC5*	0.045, 0.0273
Haller et al. 2016 [62]	Exome	391 AIS and 843 controls, targeted seq of 919 AIS	ECM (multiple), musculoskeletal collagens (multiple), *COL11A2*	6×10^{-9} (*ECM enrichment*), 1×10^{-9} (*musculoskeletal collagen enrichment*), 6×10^{-9} (*COL11A2*)
Li et al. 2016 [63]	Exome	1 family with validation in 503 controls	*AKAP2*	*Unknown*
Gao et al. 2017 [64]	Exome, linkage	1 family with targeted seq in 20 AIS families and 86 simplex patients, validation in 1038 AIS simplex and 1841 controls	*MAPK7*	2.8×10^{-5}

P-value denotes most significant published value.

during spinal surgeries. Fendri et al. conducted a microarray of IS and control osteoblasts and observed differential expression of multiple homeobox genes in IS cells [65]. Additionally, clustering analysis of differentially expressed genes showed that these genes functioned within biological pathways important in bone development. The Moreau group has also observed differences in IS osteoblasts compared to controls, including altered melatonin signaling [66, 67] and longer cilia, which they believe may affect the cell's mechanotransduction capabilities [68].

Other groups have analyzed gene expression within the paraspinal or paravertebral muscles, which extend and bend the spine and are able to be collected during spinal fusion surgery. Microarray and RT-qPCR analysis of the paraspinal muscles of IS versus control individuals revealed increased activity in TGF-β signaling, which localized to the muscle's extracellular region [69]. Paravertebral muscles were also shown to have asymmetric expression of the MT2 melatonin receptor on the convex versus concave sides of the scoliotic curve [70]. However, this finding was later disputed by Zamecnik et al., who found no difference in the expression of melatonin receptors between either the convex and concave sides of the scoliotic curve, and similarly did not find any differences between IS cases and controls [71].

Lastly, Buchan et al. analyzed genomic copy number variations (CNVs) in 148 IS patients and 1079 controls [72]. The group identified a duplication of chromosome 1q21.1 in 2.1% of IS patients, but only 0.09% of controls, as well as the presence of two chromosomal rearrangements that were previously associated with spinal phenotypes. The group concluded that over 6% of adolescent IS patients in their cohort had a clinically important copy number abnormality and suggested that copy number analysis could be clinically useful to IS patients.

Section Summary

Early genetic studies of IS were marked by specific analyses of genes or regions hypothesized to be biologically important to IS, including within the ECM, although these largely produced negative findings. Linkage analysis studies have revealed several loci associating with the disease phenotype, including SNPs in *CHD7*, although the relevance of these loci is unclear. GWAS have revealed several promising findings, with the *LBX1* and *GPR126* genes being replicated in cohorts of varying ethnicities. Other GWAS loci have not yet been replicated in other studies. The adoption of next-generation sequencing in the late 2000s also spurred several discoveries, including the identification of ECM genes as important for IS, particularly the musculoskeletal collagen genes and *HSPG2*.

Functional Studies and Animal Models

Animal modeling is an important step in research that is often required to prove that candidate genetic variations are able to cause disease. As few mammals other than humans are bipedal or develop scoliosis naturally, appropriately modeling IS has represented a significant hurdle for genetic discovery. The chicken, naturally bipedal, was used to create the first animal model of IS upon removal of the pineal gland. This phenotype was rescued upon administration of melatonin, which is secreted by the pineal gland, leading researchers to hypothesize that melatonin deficiencies may underlie IS development [73, 74]. However, subsequent linkage analysis of a region on chromosome 4q, which contained the human melatonin receptor, showed no evidence of association with IS [75]. Rats that have been made bipedal by amputation of the tail and forelimbs, coupled with gradual raising of the food and water sources, have also been used to model IS. Like the chicken, bipedalized rats likewise developed an IS-like phenotype upon pinealectomy [76, 77]. These rats were also observed to have abnormal levels of serum leptin, osteopontin, and calmodulin antagonists that were associated with spinal curve development and severity [78–81]. There is debate in the field, however, over whether these bipedal rodents accurately represent human disease or if the scoliosis is simply the result of degeneration due to an unnatural physiology.

Recently, several bony fish (teleosts) including the guppy (*Poecilia reticulata*) and zebrafish (*Danio rerio*) have emerged as leading models of IS. Scoliosis occurs naturally among several types of fish and is the most common morphological deformity [81]. These fish experience a cranial-to-caudal load, generated by swimming through water, which is hypothesized to mimic the force load experienced by humans during locomotion [81–83]. Additionally, zebrafish possess many advantages as an animal model, including rapid reproduction times, inexpensive care, ease of genetic manipulation, and abundant experimental resources including a well-annotated genome. The first fish model of IS was developed in a guppy lineage termed *curveback* [82]. The majority of the IS susceptibility of this lineage was later mapped to a 5 cM region which contained over 100 genes, including the melatonin receptor *MTNR1B* [84].

More recent studies have modeled IS in *Danio rerio* (zebrafish), a well-studied laboratory animal with a higher abundance of genetic and experimental resources as compared to the guppy. In 2014, Buchan et al. performed a forward mutagenesis screen for IS in zebrafish and isolated a recessive mutant called *skolios,* which developed an isolated spinal curvature without vertebral malformations. The phenotype was caused by a recessive mutation in *kif6*, a kinesin gene [85]. The Ciruna research group discovered that zebrafish with mutations in *ptk7* developed a late-onset spinal curvature reminiscent of adolescent IS (Fig. 7.2) [86]. The group later recreated this phenotype using temperature-sensitive mutations in multiple cilia genes with a motile cilia-specific promotor. Mutant zebrafish exhibited irregularities in cerebrospinal fluid (CSF) flow, leading the group to hypothesize that altered CSF flow may underlie the development of IS [87]. Patten et al. injected zebrafish embryos with three human *POC5* mRNA sequences identified in IS patients [40]. Injection of any of these sequences resulted in a spine deformity, without affecting other skeletal structures. The group concluded that mutations in *POC5*, which encodes a centriolar protein, may contribute to the development of IS.

Section Summary

A significant challenge in IS genetics research has been the identification of an appropriate animal model, which is often required to demonstrate disease causality of candidate genes. Pinealectomized chickens and bipedal rodents were both shown to develop scoliosis by multiple groups, although there is debate on the utility of these models. Recently, bony fishes (teleosts), particularly the zebrafish, have emerged as promising animal models. Zebrafish naturally develop spinal curvatures and may more accurately mimic scoliosis in humans due to the analogous cranial-to-caudal load experienced from swimming. Additionally, zebrafish have a well-annotated genome and fast reproduction times and are easy and inexpensive to care for. Mutations in several genes have been shown to cause an IS-like phenotype in fish models, including human *POC5* and zebrafish *ptk7*, as well as other cilia genes that lead to abnormal CSF flow.

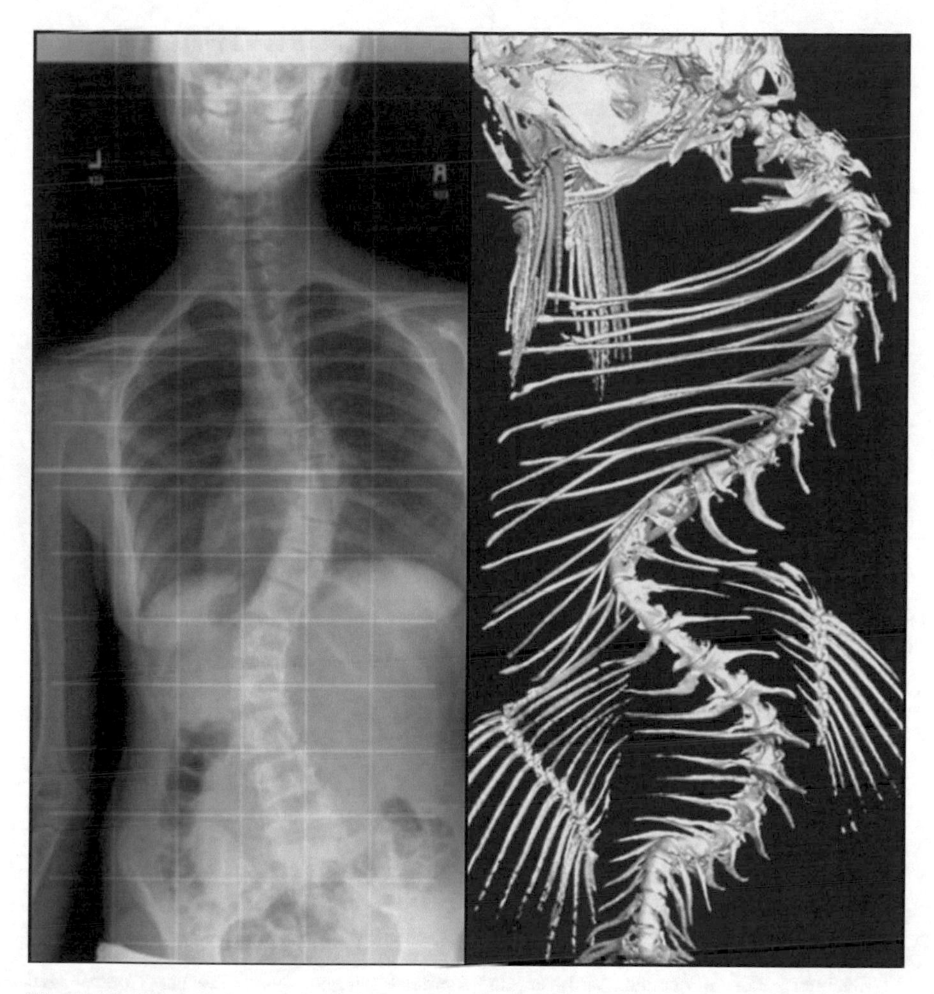

Fig. 7.2 Left, a spinal x-ray of an IS patient. Right, microcomputed tomography (micro-CT) of a *ptk7* mutant zebrafish, presenting late-onset, rotational spinal curvature mirroring defining attributes of human IS

Conclusion

Unraveling the genetic basis of IS has proven to be difficult due to extreme genetic and phenotypic heterogeneity within patient cohorts. Despite this difficulty, several candidate loci have been collectively identified by GWAS, linkage analysis, exome sequencing, and other experimental methods over the past several decades. New and emerging technologies including next-generation sequencing and CRISPR-Cas9 genetic editing present unique opportunities to discover new loci underlying IS etiology. Additionally, bony fish, particularly the zebrafish, have emerged as leading animal models to assist in demonstrating causality of candidate genomic regions in the etiology of IS. Understanding the genetic causes of IS is an important piece of the puzzle in understanding disease pathogenesis, which will help pave the way for future diagnostics, therapeutics, and perhaps cures for those affected with the disease.

Acknowledgments We thank the laboratory of Dr. Brian Ciruna for providing the micro-CT image of the mutant *ptk7* zebrafish in Fig. 7.2.

References

1. Garland HG. Hereditary scoliosis. Br Med J. 1934;1:328.
2. Wynne-Davies R. Familial (idiopathic) scoliosis. A family survey. J Bone Joint Surg Br. 1968;50(1):24–30.
3. Faber A. Untersuchungen uber die Erblichkeit der Skoliose. Arch Orthop Unfallchir. 1936;36:247–9.
4. Ma XJ, Hu P. The etiological study of idiopathic scoliosis. Zhonghua Wai Ke Za Zhi. 1994;32(8):504–6.
5. MacEwen GD, Shands AR Jr. Scoliosis--a deforming childhood problem. Clin Pediatr (Phila). 1967;6(4):210–6.
6. Perricone G, Paradiso T. Familial factors in so-called idiopathic scoliosis. Chir Organi Mov. 1987;72(4):355–8.
7. De George FV, Fisher RL. Idiopathic scoliosis: genetic and environmental aspects. J Med Genet. 1967;4(4):251–7.
8. Riseborough EJ, Wynne-Davies R. A genetic survey of idiopathic scoliosis in Boston, Massachusetts. J Bone Joint Surg Am. 1973;55(5):974–82.
9. Bonaiti C, Feingold J, Briard ML, Lapeyre F, Rigault P, Guivarch J. Genetics of idiopathic scoliosis. Helv Paediatr Acta. 1976;31(3):229–40.
10. Ogilvie JW, Braun J, Argyle V, Nelson L, Meade M, Ward K. The search for idiopathic scoliosis genes. Spine. 2006;31(6):679–81.
11. Carr AJ. Adolescent idiopathic scoliosis in identical twins. J Bone Joint Surg Br. 1990;72(6:1077.
12. Gaertner RL. Idiopathic scoliosis in identical (monozygotic) twins. South Med J. 1979;72(2):231–4.
13. Senda M, Harada Y, Nakahara S, Inoue H. Lumbar spinal changes over 20 years after posterior fusion for idiopathic scoliosis. Acta Med Okayama. 1997;51(6):327–31.
14. Kesling KL, Reinker KA. Scoliosis in twins. A meta-analysis of the literature and report of six cases. Spine. 1997;22(17):2009–14. discussion 15
15. McKinley LM, Leatherman KD. Idiopathic and congenital scoliosis in twins. Spine. 1978;3(3):227–9.
16. Murdoch G. Scoliosis in twins. J Bone Joint Surg Br. 1959;41-B:736–7.
17. Scott TF, Bailey RW. Idiopathic scoliosis in fraternal twins. J Mich State Med Soc. 1963;62:283–4.
18. Cowell HR, Hall JN, MacEwen GD. Genetic aspects of idiopathic scoliosis. A Nicholas Andry award essay, 1970. Clin Orthop Relat Res. 1972;86:121–31.
19. Miller NH, Schwab DL, Sponseller P, Shugert E, Bell J, Maestri N. Genomic search for X-linkage in familial adolescent idiopathic scoliosis. In: IRSoSDMn, editor. Research into spinal deformities 2. Amsterdam: IOS Press; 1998. p. 209–13.
20. Czeizel A, Bellyei A, Barta O, Magda T, Molnar L. Genetics of adolescent idiopathic scoliosis. J Med Genet. 1978;15(6):424–7.

21. Kruse LM, Buchan JG, Gurnett CA, Dobbs MB. Polygenic threshold model with sex dimorphism in adolescent idiopathic scoliosis: the Carter effect. J Bone Joint Surg Am. 2012;94(16):1485–91.
22. Robin GC, Cohen T. Familial scoliosis. A clinical report. J Bone Joint Surg Br. 1975;57(2):146–8.
23. Carr AJ, Ogilvie DJ, Wordsworth BP, Priestly LM, Smith R, Sykes B. Segregation of structural collagen genes in adolescent idiopathic scoliosis. Clin Orthop Relat Res. 1992;(274):305–10.
24. Bell M, Teebi AS. Autosomal dominant idiopathic scoliosis? Am J Med Genet. 1995;55(1):112.
25. Miller NH, Mims B, Child A, Milewicz DM, Sponseller P, Blanton SH. Genetic analysis of structural elastic fiber and collagen genes in familial adolescent idiopathic scoliosis. J Orthop Res. 1996;14(6):994–9.
26. Aksenovich TI, Semenov IR, Ginzburg E, Zaidman AM. Preliminary analysis of inheritance of scoliosis. Genetika. 1988;24(11):2056–63.
27. Axenovich TI, Zaidman AM, Zorkoltseva IV, Tregubova IL, Borodin PM. Segregation analysis of idiopathic scoliosis: demonstration of a major gene effect. Am J Med Genet. 1999;86(4):389–94.
28. Justice CM, Miller NH, Marosy B, Zhang J, Wilson AF. Familial idiopathic scoliosis: evidence of an X-linked susceptibility locus. Spine. 2003;28(6):589–94.
29. Berquet KH. Considerations on heredity in idiopathic scoliosis. Z Orthop Ihre Grenzgeb. 1966;101(2):197–209.
30. Shapiro JR, Burn VE, Chipman SD, Velis KP, Bansal M. Osteoporosis and familial idiopathic scoliosis: association with an abnormal alpha 2(I) collagen. Connect Tissue Res. 1989;21(1–4):117–23. discussion 24
31. Levaia NV. Genetic aspect of dysplastic (idiopathic) scoliosis. Ortop Travmatol Protez. 1981;(2):23–9.
32. Marosy B, Justice CM, Nzegwu N, Kumar G, Wilson AF, Miller NH. Lack of association between the aggrecan gene and familial idiopathic scoliosis. Spine. 2006;31(13):1420–5.
33. Wise CA, Barnes R, Gillum J, Herring JA, Bowcock AM, Lovett M. Localization of susceptibility to familial idiopathic scoliosis. Spine. 2000;25(18):2372–80.
34. Chan V, Fong GC, Luk KD, Yip B, Lee MK, Wong MS, et al. A genetic locus for adolescent idiopathic scoliosis linked to chromosome 19p13.3. Am J Hum Genet. 2002;71(2):401–6.
35. Salehi LB, Mangino M, De Serio S, De Cicco D, Capon F, Semprini S, et al. Assignment of a locus for autosomal dominant idiopathic scoliosis (IS) to human chromosome 17p11. Hum Genet. 2002;111(4–5):401–4.
36. Miller NH, Justice CM, Marosy B, Doheny KF, Pugh E, Zhang J, et al. Identification of candidate regions for familial idiopathic scoliosis. Spine. 2005;30(10):1181–7.
37. Gao X, Gordon D, Zhang D, Browne R, Helms C, Gillum J, et al. CHD7 gene polymorphisms are associated with susceptibility to idiopathic scoliosis. Am J Hum Genet. 2007;80(5):957–65.
38. Tilley MK, Justice CM, Swindle K, Marosy B, Wilson AF, Miller NH. CHD7 gene polymorphisms and familial idiopathic scoliosis. Spine (Phila Pa 1976). 2013;38(22):E1432–6.
39. Edery P, Margaritte-Jeannin P, Biot B, Labalme A, Bernard JC, Chastang J, et al. New disease gene location and high genetic heterogeneity in idiopathic scoliosis. Eur J Hum Genet. 2011;19(8):865–9.
40. Patten SA, Margaritte-Jeannin P, Bernard JC, Alix E, Labalme A, Besson A, et al. Functional variants of POC5 identified in patients with idiopathic scoliosis. J Clin Invest. 2015;125(3):1124–8.
41. Sharma S, Gao X, Londono D, Devroy SE, Mauldin KN, Frankel JT, et al. Genome-wide association studies of adolescent idiopathic scoliosis suggest candidate susceptibility genes. Hum Mol Genet. 2011;20(7):1456–66.
42. Takahashi Y, Kou I, Takahashi A, Johnson TA, Kono K, Kawakami N, et al. A genome-wide association study identifies common variants near LBX1 associated with adolescent idiopathic scoliosis. Nat Genet. 2011;43(12):1237–40.
43. Fan YH, Song YQ, Chan D, Takahashi Y, Ikegawa S, Matsumoto M, et al. SNP rs11190870 near LBX1 is associated with adolescent idiopathic scoliosis in southern Chinese. J Hum Genet. 2012;57:244–6.

44. Gao W, Peng Y, Liang G, Liang A, Ye W, Zhang L, et al. Association between common variants near LBX1 and adolescent idiopathic scoliosis replicated in the Chinese Han population. PLoS One. 2013;8(1):e53234.
45. Jiang H, Qiu X, Dai J, Yan H, Zhu Z, Qian B, et al. Association of rs11190870 near LBX1 with adolescent idiopathic scoliosis susceptibility in a Han Chinese population. Eur Spine J. 2013;22(2):282–6.
46. Londono D, Kou I, Johnson TA, Sharma S, Ogura Y, Tsunoda T, et al. A meta-analysis identifies adolescent idiopathic scoliosis association with LBX1 locus in multiple ethnic groups. J Med Genet. 2014;51(6):401–6.
47. Grauers A, Wang J, Einarsdottir E, Simony A, Danielsson A, Akesson K, et al. Candidate gene analysis and exome sequencing confirm LBX1 as a susceptibility gene for idiopathic scoliosis. Spine J. 2015;15(10):2239–46.
48. Zhu Z, Tang NL, Xu L, Qin X, Mao S, Song Y, et al. Genome-wide association study identifies new susceptibility loci for adolescent idiopathic scoliosis in Chinese girls. Nat Commun. 2015;6:8355.
49. Chettier R, Nelson L, Ogilvie JW, Albertsen HM, Ward K. Haplotypes at LBX1 have distinct inheritance patterns with opposite effects in adolescent idiopathic scoliosis. PLoS One. 2015;10(2):e0117708.
50. Cao Y, Min J, Zhang Q, Li H, Li H. Associations of LBX1 gene and adolescent idiopathic scoliosis susceptibility: a meta-analysis based on 34,626 subjects. BMC Musculoskelet Disord. 2016;17:309.
51. Liu S, Wu N, Zuo Y, Zhou Y, Liu J, Liu Z, et al. Genetic polymorphism of LBX1 is associated with adolescent idiopathic scoliosis in northern Chinese Han population. Spine (Phila Pa 1976). 2017;42:1125–9.
52. Nada D, Julien C, Samuels ME, Moreau A. A replication study for association of LBX1 locus with adolescent idiopathic scoliosis in French-Canadian population. Spine (Phila Pa 1976). 2017;43:172–178.
53. Kou I, Takahashi Y, Johnson TA, Takahashi A, Guo L, Dai J, et al. Genetic variants in GPR126 are associated with adolescent idiopathic scoliosis. Nat Genet. 2013;45(6):676–9.
54. Xu JF, Yang GH, Pan XH, Zhang SJ, Zhao C, Qiu BS, et al. Association of GPR126 gene polymorphism with adolescent idiopathic scoliosis in Chinese populations. Genomics. 2015;105(2):101–7.
55. Qin X, Xu L, Xia C, Zhu W, Sun W, Liu Z, et al. Genetic variant of GPR126 gene is functionally associated with adolescent idiopathic scoliosis in Chinese population. Spine (Phila Pa 1976). 2017;42:E1098–103.
56. Ogura Y, Kou I, Miura S, Takahashi A, Xu L, Takeda K, et al. A functional SNP in BNC2 is associated with adolescent idiopathic scoliosis. Am J Hum Genet. 2015;97(2):337–42.
57. Miyake A, Kou I, Takahashi Y, Johnson TA, Ogura Y, Dai J, et al. Identification of a susceptibility locus for severe adolescent idiopathic scoliosis on chromosome 17q24.3. PLoS One. 2013;8(9):e72802.
58. Sharma S, Londono D, Eckalbar WL, Gao X, Zhang D, Mauldin K, et al. A PAX1 enhancer locus is associated with susceptibility to idiopathic scoliosis in females. Nat Commun. 2015;6:6452.
59. Zhu Z, Xu L, Leung-Sang Tang N, Qin X, Feng Z, Sun W, et al. Genome-wide association study identifies novel susceptible loci and highlights Wnt/beta-catenin pathway in the development of adolescent idiopathic scoliosis. Hum Mol Genet. 2017;26(8):1577–83.
60. Buchan JG, Alvarado DM, Haller GE, Cruchaga C, Harms MB, Zhang T, et al. Rare variants in FBN1 and FBN2 are associated with severe adolescent idiopathic scoliosis. Hum Mol Genet. 2014;23:5271–82.
61. Baschal EE, Wethey CI, Swindle K, Baschal RM, Gowan K, Tang NL, et al. Exome sequencing identifies a rare HSPG2 variant associated with familial idiopathic scoliosis. G3 (Bethesda). 2014;5(2):167–74.

62. Haller G, Alvarado D, McCall K, Yang P, Cruchaga C, Harms M, et al. A polygenic burden of rare variants across extracellular matrix genes among individuals with adolescent idiopathic scoliosis. Hum Mol Genet. 2016;25(1):202–9.
63. Li W, Li Y, Zhang L, Guo H, Tian D, Li Y, et al. AKAP2 identified as a novel gene mutated in a Chinese family with adolescent idiopathic scoliosis. J Med Genet. 2016;53(7):488–93.
64. Gao W, Chen C, Zhou T, Yang S, Gao B, Zhou H, et al. Rare coding variants in MAPK7 predispose to adolescent idiopathic scoliosis. Hum Mutat. 2017;38(11):1500–10.
65. Fendri K, Patten SA, Kaufman GN, Zaouter C, Parent S, Grimard G, et al. Microarray expression profiling identifies genes with altered expression in adolescent idiopathic scoliosis. Eur Spine J. 2013;22(6):1300–11.
66. Moreau A, Wang DS, Forget S, Azeddine B, Angeloni D, Fraschini F, et al. Melatonin signaling dysfunction in adolescent idiopathic scoliosis. Spine (Phila Pa 1976). 2004;29(16):1772–81.
67. Azeddine B, Letellier K, Wang da S, Moldovan F, Moreau A. Molecular determinants of melatonin signaling dysfunction in adolescent idiopathic scoliosis. Clin Orthop Relat Res. 2007;462:45–52.
68. Oliazadeh N, Gorman KF, Eveleigh R, Bourque G, Moreau A. Identification of elongated primary cilia with impaired Mechanotransduction in idiopathic scoliosis patients. Sci Rep. 2017;7:44260.
69. Nowak R, Kwiecien M, Tkacz M, Mazurek U. Transforming growth factor-beta (TGF- beta) signaling in paravertebral muscles in juvenile and adolescent idiopathic scoliosis. Biomed Res Int. 2014;2014:594287.
70. Qiu Y, Wu L, Wang B, Yu Y, Zhu Z. Asymmetric expression of melatonin receptor mRNA in bilateral paravertebral muscles in adolescent idiopathic scoliosis. Spine (Phila Pa 1976). 2007;32(6):667–72.
71. Zamecnik J, Krskova L, Hacek J, Stetkarova I, Krbec M. Etiopathogenesis of adolescent idiopathic scoliosis: expression of melatonin receptors 1A/1B, calmodulin and estrogen receptor 2 in deep paravertebral muscles revisited. Mol Med Rep. 2016;14(6):5719–24.
72. Buchan JG, Alvarado DM, Haller G, Aferol H, Miller NH, Dobbs MB, et al. Are copy number variants associated with adolescent idiopathic scoliosis? Clin Orthop Relat Res. 2014;472(10):3216–25.
73. Machida M, Dubousset J, Imamura Y, Iwaya T, Yamada T, Kimura J. Role of melatonin deficiency in the development of scoliosis in pinealectomised chickens. J Bone Joint Surg Br. 1995;77(1):134–8.
74. Bagnall K, Raso VJ, Moreau M, Mahood J, Wang X, Zhao J. The effects of melatonin therapy on the development of scoliosis after pinealectomy in the chicken. J Bone Joint Surg Am. 1999;81(2):191–9.
75. Morcuende JA, Minhas R, Dolan L, Stevens J, Beck J, Wang K, et al. Allelic variants of human melatonin 1A receptor in patients with familial adolescent idiopathic scoliosis. Spine. 2003;28(17):2025–8. discussion 9
76. Machida M, Murai I, Miyashita Y, Dubousset J, Yamada T, Kimura J. Pathogenesis of idiopathic scoliosis. Experimental study in rats. Spine (Phila Pa 1976). 1999;24(19):1985–9.
77. Machida M, Saito M, Dubousset J, Yamada T, Kimura J, Shibasaki K. Pathological mechanism of idiopathic scoliosis: experimental scoliosis in pinealectomized rats. Eur Spine J. 2005;14(9):843–8.
78. Akel I, Kocak O, Bozkurt G, Alanay A, Marcucio R, Acaroglu E. The effect of calmodulin antagonists on experimental scoliosis: a pinealectomized chicken model. Spine (Phila Pa 1976). 2009;34(6):533–8.
79. Wu T, Sun X, Zhu Z, Zheng X, Qian B, Zhu F, et al. Role of high central leptin activity in a scoliosis model created in bipedal amputated mice. Stud Health Technol Inform. 2012;176:31–5.
80. Yadav MC, Huesa C, Narisawa S, Hoylaerts MF, Moreau A, Farquharson C, et al. Ablation of Osteopontin improves the skeletal phenotype of Phospho1 mice. J Bone Miner Res. 2014;29:2369–81.

81. Boswell CW, Ciruna B. Understanding idiopathic scoliosis: a new zebrafish School of Thought. Trends Genet. 2017;33(3):183–96.
82. Gorman KF, Breden F. Teleosts as models for human vertebral stability and deformity. Comp Biochem Physiol C Toxicol Pharmacol. 2007;145(1):28–38.
83. Gorman KF, Breden F. Idiopathic-type scoliosis is not exclusive to bipedalism. Med Hypotheses. 2009;72(3):348–52.
84. Gorman KF, Christians JK, Parent J, Ahmadi R, Weigel D, Dreyer C, et al. A major QTL controls susceptibility to spinal curvature in the curveback guppy. BMC Genet. 2011;12:16.
85. Buchan JG, Gray RS, Gansner JM, Alvarado DM, Burgert L, Gitlin JD, et al. Kinesin family member 6 (kif6) is necessary for spine development in zebrafish. Dev Dyn. 2014;243(12):1646–57.
86. Hayes M, Gao X, Yu LX, Paria N, Henkelman RM, Wise CA, et al. ptk7 mutant zebrafish models of congenital and idiopathic scoliosis implicate dysregulated Wnt signalling in disease. Nat Commun. 2014;5:4777.
87. Grimes DT, Boswell CW, Morante NF, Henkelman RM, Burdine RD, Ciruna B. Zebrafish models of idiopathic scoliosis link cerebrospinal fluid flow defects to spine curvature. Science. 2016;352(6291):1341–4.
88. Inoue M, Minami S, Nakata Y, Kitahara H, Otsuka Y, Isobe K, et al. Association between estrogen receptor gene polymorphisms and curve severity of idiopathic scoliosis. Spine. 2002;27(21):2357–62.
89. Alden KJ, Marosy B, Nzegwu N, Justice CM, Wilson AF, Miller NH. Idiopathic scoliosis: identification of candidate regions on chromosome 19p13. Spine. 2006;31(16):1815–9.
90. Yeung HY, Tang NL, Lee KM, Ng BK, Hung VW, Kwok R, et al. Genetic association study of insulin-like growth factor-I (IGF-I) gene with curve severity and osteopenia in adolescent idiopathic scoliosis. Stud Health Technol Inform. 2006;123:18–24.
91. Wu J, Qiu Y, Zhang L, Sun Q, Qiu X, He Y. Association of estrogen receptor gene polymorphisms with susceptibility to adolescent idiopathic scoliosis. Spine. 2006;31(10):1131–6.
92. Tang NL, Yeung HY, Lee KM, Hung VW, Cheung CS, Ng BK, et al. A relook into the association of the estrogen receptor [alpha] gene (PvuII, XbaI) and adolescent idiopathic scoliosis: a study of 540 Chinese cases. Spine. 2006;31(21):2463–8.
93. Qiu XS, Tang NL, Yeung HY, Qiu Y, Qin L, Lee KM, et al. The role of melatonin receptor 1B gene (MTNR1B) in adolescent idiopathic scoliosis--a genetic association study. Stud Health Technol Inform. 2006;123:3–8.
94. Montanaro L, Parisini P, Greggi T, Di Silvestre M, Campoccia D, Rizzi S, et al. Evidence of a linkage between matrilin-1 gene (MATN1) and idiopathic scoliosis. Scoliosis. 2006;1:21.
95. Ocaka L, Zhao C, Reed JA, Ebenezer ND, Brice G, Morley T, et al. Assignment of two loci for autosomal dominant adolescent idiopathic scoliosis to chromosomes 9q31.2-q34.2 and 17q25.3-qtel. J Med Genet. 2008;45(2):87–92.
96. Raggio CL, Giampietro PF, Dobrin S, Zhao C, Dorshorst D, Ghebranious N, et al. A novel locus for adolescent idiopathic scoliosis on chromosome 12p. J Orthop Res. 2009;27(10):1366–72.
97. Marosy B, Justice C, Vu C, Zorn A, Nzegwu N, Wilson A, et al. Identification of susceptibility loci for scoliosis in FIS families with triple curves. AJMG. 2010;152A:846–55.
98. Clough M, Justice CM, Marosy B, Miller NH. Males with familial idiopathic scoliosis: a distinct phenotypic subgroup. Spine (Phila Pa 1976). 2010;35(2):162–8.

Chapter 8
Adolescence and Scoliosis: Deciphering the Complex Biology of Puberty and Scoliosis

Jeremy McCallum-Loudeac and Megan J. Wilson

Introduction

Scoliosis is a disorder with large heterogeneity in onset, presentation and progression of spinal curvature. Scoliotic curvatures are generally classed into two forms: (a) congenital, whereby an inherited disorder causes scoliotic curvature through incorrect development or the vertebrae or vertebrae-associated structures, and (b) idiopathic, arising spontaneously during growth and development, typically with no physical defect to the bony elements of the vertebrae and with little understanding of aetiopathogenesis [29]. Idiopathic forms of the disease are far more prevalent, the largest proportion of these arising during adolescence, between 10 and 18 years of age; this is termed adolescent idiopathic scoliosis (AIS) [13, 28]. The onset and progression of AIS is poorly understood, making early intervention difficult and preventing development of a diagnostic tool for predicting curve progression. Treatment options include invasive surgery and/or thoracic bracing, often for up to 22 h a day [11].

One of the most striking features of AIS biology is the sex bias observed in patient groups; females are significantly more affected and are more likely to progress to a severe curvature [28]. The sex ratio of AIS is 1.6:1 female-male at 10 years of age, increasing to 6.4:1 at 11 years of age. Furthermore, the observed prevalence is equal between sexes with curvatures of 10°, whereas for curvatures of at least 30°, a ratio of ten females for every one male is observed [76]. Substantial heterogeneity is seen in the likelihood and extent of curve progression between males and females, with factors such as age, pubertal stage of growth, original curve type and angle and events such as menses onset or peak growth periods all impacting on curve progression [28].

J. McCallum-Loudeac · M. J. Wilson (✉)
Department of Anatomy, University of Otago, Dunedin, New Zealand
e-mail: mccje470@student.otago.ac.nz; meganj.wilson@otago.ac.nz

K. Kusumi, S. L. Dunwoodie (eds.), *The Genetics and Development of Scoliosis*, https://doi.org/10.1007/978-3-319-90149-7_8

Given the complexity of biological changes that occur during adolescence, it is possible that AIS has multiple origins resulting in presentation of a similar phenotype. Puberty involves the integration of multiple molecular pathways, resulting in the adolescent growth spurt, along with reproductive maturation. Many of these pathways have sex-dimorphic characteristics, such as differing expression levels between boys and girls during puberty, and have been implicated in AIS biology. Once we have a better understanding why there is a sex bias, we will have a better understanding of not only the disease itself but drivers of spinal curve progression. For diseases developing in the adolescent period, particularly those resulting from aberrant growth, understanding the processes underlying these changes will provide better understanding of disease onset and progression.

Puberty and Spine Biology

Puberty Overview

Puberty, marking the end of childhood and the onset of adulthood, is characterised by attainment of sexual maturation and the development of secondary sexual characteristics, as well as changes to overall body size and shape. Fluctuating hormone levels over the peri-, pre-, pubertal and post-pubertal stages are responsible for many anatomical and behavioural changes experienced throughout adolescence [1].

Puberty onset is typically between 11 and 13 years of age; however, in some populations, there can be up to a 5-year variation in onset, believed to be attributable to industrialisation, nutritional status, genetics, ethnicity and sex [1]. Females will undergo and complete puberty earlier ~2 years earlier than males [3]. The age of menarche is determined by genetic, hormonal and environmental factors and is considered an important biological event, as it will dictate lifelong exposure to oestrogen, a factor critically important for health outcomes later in life. Marshall and Tanner [46] classified puberty into five stages based on the changes to the breasts, pubic hair and genital development, a scale which has been widely used since [46].

Pubertal onset and progression is largely controlled through the hypothalamic-pituitary-gonadal (HPG) axis, with onset attributed to changes in the secretory pattern of gonadotrophin-releasing hormone (GnRH). The HPG axis is dormant during the prepubertal period, until an increase in GnRH secretion by neurons located in the hypothalamus stimulates the pituitary gland to produce luteinizing hormone (LH) and follicle-stimulating hormone (FSH), and in turn, they stimulate production of the gonadal steroid hormones (Fig. 8.1) (reviewed in [23]).

Although the process of sexual maturation is most commonly associated with puberty, there is also a noteworthy growth spurt resulting in increased vertical height. This growth spurt is largely restricted to the thorax; two-thirds of the overall height increase over this period is attributed to trunk lengthening [19]. This pubertal

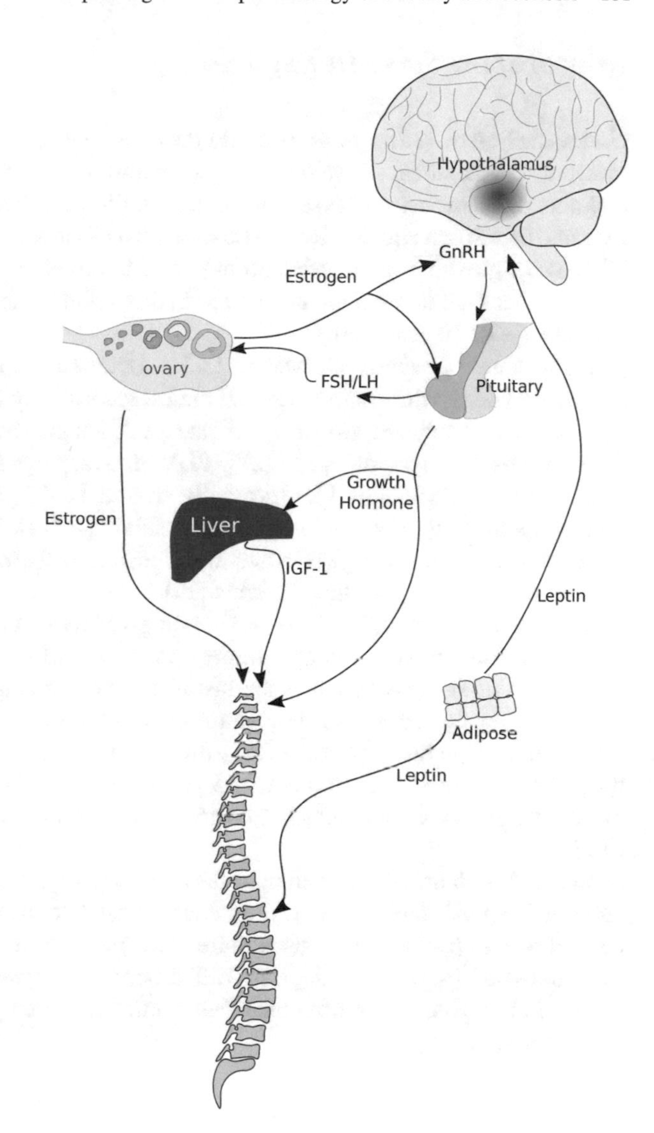

Fig. 8.1 Overview of the signalling pathways involved in regulating female puberty and have been implicated in AIS biology. At the onset of puberty, pulses of GnRH secretion from the hypothalamus stimulate the pituitary gland to produce FSH and LH. These two hormones activate ovarian oestrogen synthesis, leading to a rise in circulating oestrogen levels. Leptin released by adipose tissue communicates to the hypothalamus to maintain body weight and energy homeostasis. Growth hormone production also increases during puberty; this increases production of IGF-1 by the liver. Together these hormones are required for sexual maturation and the adolescent growth spurt, including increased growth of the spinal column. Components of all of these pathways have been implicated in AIS biology through genetic and cellular biology studies

adolescent growth spurt is a large period of physical change, and although well studied, its onset is hard to predict [3]. The adolescent growth spurt is commonly observed in females 12 months prior to menarche, while in males it follows the onset of puberty and may be up to 2 years later than females on average [74]. This is also the period where boys overtake girls in overall height.

Growth of the Spine During Puberty

Understanding disorders of growth and development requires an understanding of 'normal' development. Development and maturation of the spine begins in utero and continues through to the second decade of life [10]. Puberty is a critical period for bone deposition and density; ~40% of bone mass is acquired during adolescence [62]. Spine growth over the adolescent period is driven by an increase in vertebral height rather than disc height; no change in disc height was observed in individuals over the age of 10 years [69].

At the onset of puberty, an average 11% (~18 cm) and 13% (~20 cm) of standing height growth remains for females and males, respectively [18]. The period of time with the fastest upward growth is termed peak height velocity (PHV), timing of which varies with pubertal stage [22]. PHV also displays sex-specific differences, with females experiencing PHV typically around 11–13 years of skeletal age and males experiencing it at 13–15 years of skeletal age [12]. During the peak growth period, boys will overtake girls in overall height and will eventually continue growth for a further 2 years more than females [19].

The thoracic spine (T1–T12) will undergo several rapid growth periods: the first being from birth to 5 years of age, an increase of ~7 cm; the second is from 5–10 years of age (~4 cm); and the last is another rapid growth through puberty (~7 cm) over the course of 3–4 years [19]. The lumbar spine will follow the same trend as the thoracic spine for growth periods, initially growing 3 cm in the first 5 years of life, 2 cm in the next 5 years and 3 cm during puberty. The spine has an incredibly slow process of ossification, finishing only in the 25th year of life [19].

Support structures associated with the bony elements of the spine play an important role in spinal stability. In the prepubertal period, females have been observed to have 10% less paraspinous musculature than males, a trend observed to continue into adulthood [4, 42]. Although no differences were observed between sexes for muscle fibre types, minor differences were noted based on physical exercise during early development [42].

AIS Onset and Progression

It is important to distinguish between curvature onset and curve progression during adolescence, as recent genetic studies indicate that some sequence variants may be associated with scoliosis onset, whereas others are linked to progression risk. This section will briefly look at both onset and curve progression in adolescence and subsequently, the factors believed to influence curve progression.

Age of Onset

Scoliosis first observed in the adolescent period, and in anyone diagnosed over the age of 10 years is typically classified into the cohort of AIS. Studies specifically reporting ages of onset between sexes are uncommon, with the true age of onset difficult to determine as curves are often not noticed until some progression has occurred. Moreover, clinicians typically look at markers of sexual or skeletal maturity with the aim of providing some prognosis to the patient, and these are the most common end points to report [81]. Accurately determining the onset of scoliosis would allow for further work into aetiopathogenesis, allowing clinicians and researchers to understand the initial events leading to scoliotic onset. Several studies over the last few decades have provided some information regarding age of onset in their cohorts [2, 82]. Wynne-Davies [82] reported the age of onset in a small Scottish cohort, finding that the most prevalent onset for females was 14 years and 15 years of age for males [82]. Adobor et al. [2] reported the mean age of onset in their Norwegian cohort, 14.5 ± 2.1 and 15.5 ± 2.1 years for girls and boys, respectively [2].

Curve Progression

Curve degree and likelihood of progression in AIS patients is considered an important factor in determining appropriate treatment options and, as such, is a highly recorded and studied aspect of the disease. Understanding curve progression rates at the different stages of pubertal/sexual maturation provides good insight into the likelihood or curve progression and which processes may be associated with advanced curve progression.

Earlier AIS onset (<12 years of age) and a larger spinal curve that presents earlier in adolescence are considered high risk of curve progression, particularly during the peak height velocity period, whereas the older the individual at first presentation with AIS, the lower the incidence of progression [37, 88]. Curve progression during adolescence also plays a role in the continued progression of scoliotic curves following skeletal maturity; the larger the curve at skeletal maturity, the greater the progression [80]. Ylikoski [88] reports scoliotic progression to be dependent on several factors: growth velocity, age at presentation (first clinical visit), Risser sign, skeletal maturity and sexual maturity. The most significant curve progression was exhibited between 9 and 13 years of age, with an average growth velocity of 2 cm/year, in patients with bone ages between 9 and 14 years and Risser signs of 0–1 and typically 0.5–2 years before menarche [88].

Puberty-Associated Indicators of Progression Risk

Factors influencing curve progression are poorly understood. Sex-specific patterns of growth, differences in age at growth spurt, sexual maturity and skeletal maturity have all been investigated as possible predictors of curve progression. Several aspects common among progressive scoliotic patients include a lack of pubertal development (lower Tanner stage, 0–2), low Risser sign (0–1) and a large spinal curvature at initial presentation (30–45°). Alternative factors contributing to curve progression include the curve type (single, double, thoracic or lumbar), location of the curve apex and the presence of any rotation in the apical vertebra [80].

Current data regarding age of menarche and scoliosis onset (or progression) is conflicting, with mixed results. Part of the reason for these conflicting results is the lack of reliable and consistent data in historical studies regarding puberty and abnormal growth contributing to AIS. Studying the pubertal transition on a population scale is difficult due to the large heterogeneity between study designs and an inability to confidently (or accurately) use one, or more, body characteristics as a marker of pubertal onset. Mao et al. [43] compiled data from ~6000 healthy female adolescents and ~2200 females presenting with AIS, noting a delayed onset of menarche in those with AIS [43]. Differences in menarche age between the groups, although statistically significant, was slight, with control girls undergoing menarche at 12.63 years of age and AIS girls undergoing menarche at 12.83 years of age [43].

The PHV period is typically considered the period where the scoliotic curvatures are most likely to arise [33]. This growth spurt appears to occur at a more rapid rate in many scoliotic individuals, with a 13–30% greater increase in height over the same time period as adolescents who never developed scoliosis, even though they might reach a similar final height to non-scoliotic children. This suggests that progression of AIS is linked to rapid growth [36, 65]. However, it is difficult to predict when PHV will occur even with respect to other puberty markers such as stages of sexual maturation [22].

Unlike females, few studies have been conducted looking at the anthropomorphic measurements of male AIS patients [77]. Wang [77] report standing and sitting heights, following correction, were all comparable between patients and controls in the male population [77]. These findings suggest female-specific processes responsible for the significant increase in height of females over the pubertal stages; still what drives these processes leading to such a drastic sex bias in AIS is unknown.

Why might it be important to examine and understand the differences in pubertal development between male and females? As discussed further below, circulating hormones and their downstream signalling pathways influence adolescence development in a sex-dimorphic fashion, an important factor that may contribute to AIS and the onset of puberty, particularly in females. This has an important bearing on the timing of onset and progression of spine curvatures. Additionally, it is difficult to determine whether changes to the anatomical aspects such as the skeleton are causal or a consequence of spinal curvature.

Puberty-Associated Factors Influencing Curve Progression

Recent use of genome-wide association studies has provided a large number of common variants for the onset and/or susceptibility of AIS. Many of the variants associated with AIS onset and development are also associated with regulation of puberty (onset and progression). The factors believed to result in scoliotic curvatures will be presented here, with their role in pubertal regulation discussed below. Single-nucleotide polymorphisms associated with AIS have been identified in a variety of genes: oestrogen receptors (*ESR1* and *ESR2*), melatonin receptors (*MNTR1A* and *MNTR1B*), insulin-like growth factor 1 (*IGF-1*) and the leptin receptor (*Ob-R*) [24, 35, 53, 54, 59, 87, 89]. However not all of these factors have been associated with AIS across multiple ethnic groups, warranting further investigation into AIS genetics and biology.

Oestrogen and Oestrogen Receptors

The synthesis and production of oestrogen is critical in pubertal maturation and proper sexual functioning in both males and females. The majority of oestrogens are produced by the gonads in females (Fig. 8.1), with the remainder formed from testosterone aromatisation [66]. Oestrogen levels in females are closely related to the maturation of sexual organs and an increase in testosterone through life [16]. In males, oestradiol, largely produced by aromatisation of testosterone in peripheral adipose tissue, maintains testis Leydig cell populations [48].

The common form of circulating oestrogen is 17β-oestradiol (E2), which acts through binding to the intracellular oestrogen receptors ERα and ERβ, encoded by the *Esr1* and *Esr2* genes, respectively [86]. Oestrogen receptors not only mediate oestrogen response in reproductive tissues but are found throughout many tissues in the body including the CNS and bone [86]. These ligand-activated transcription factors translocate to the nucleus to bind to oestrogen response elements (EREs) to oestrogens genomic actions. Oestrogens also induce non-genomic actions, through an intracellular G-protein-coupled oestrogen receptor (GPER), although the mechanism is not as well understood [64]. GPER is also expressed in multiple tissues including skeletal muscle and the CNS [58] and mediates rapid signalling events in response to oestrogen [63].

Clinical studies measuring circulating steroid hormone levels in AIS patients have produced contradictory findings, leading to suggestions that oestrogen role in AIS pathogenesis is likely due to altered responses to oestrogen levels, rather the level of oestrogen itself [30]. Although the literature lacks substantial evidence implicating E2 levels as a direct cause for scoliosis in humans, several animal model studies indicate that oestrogen exposure can alter scoliosis incidence and severity [17, 25].

Human genetic studies have focused on associations between AIS risk and ER polymorphisms, with mixed results. Two gene polymorphisms in particular, rs9340799 (XbaI) and rs2234693 (PvuII), have been examined in different ethnic cohorts. Inoue et al. [24] were the first to show an association between oestrogen receptor polymorphisms and curve severity [24]. Initially, Inoue [24] showed that the XbaI polymorphism was associated with curve severity and progression but not onset [24]. The SNP rs9340799 was also linked to susceptibility for AIS and female bias, while the second SNP (rs2234693/XbaI) appeared to have no significant risk association. However, these findings were not replicated by further studies in Chinese AIS patients [73]. Contrasting the findings of Inoue et al. [24], Nikolova and colleagues identified the SNP rs2234693 (PvuII) as associated with disease susceptibility and curve severity among a Bulgarian cohort [53].

Janusz et al. [26] investigated a cohort of female patients presenting with AIS to determine whether or not association between the *ESR1* and *ESR2* gene polymorphisms and age at menarche is a potential cause of AIS. The age of menarche of this study population did not significantly differ from the general population. Furthermore, there was no difference between genotype of patients and the age at menarche [26]. The polymorphisms in *ESR1* and *ESR2* did not show any association with age at menarche, however in this cohort, girls with severe AIS exhibited a delayed onset of menarche [26].

A review of the oestrogen receptor studies conducted by Yang et al. [83] lead the authors to conclude there is not yet sufficient evidence to link *ESR1* gene polymorphisms to onset of AIS [83]. The differences between studies vary too greatly to draw any sound conclusions. However, they predict that the *ESR1* gene is potential modifier of AIS curve progression.

A recent review of the association between *ESR2* polymorphisms (rs1256120) and AIS was conducted by Zhao et al. [90]; three studies were included comprising of Japanese, Chinese and Caucasian cohorts [90]. Though a single association for *ESR2* and AIS was observed in the Chinese cohort, subsequent replication studies have not shown additional associations, leading authors to conclude that *ESR2* polymorphisms are neither associated with AIS susceptibility nor curve progression [90]. Nonetheless, a third oestrogen receptor has also been linked to AIS, with *GPER* gene polymorphisms (rs3808351, rs10269151 and rs426655s3) significantly associated with curve severity in AIS patients [57].

Melatonin Pathway

Melatonin is a hormone secreted by the pineal gland, with important roles in the regulation of circadian rhythm, sexual maturation, ageing and bone structure [45, 56]. Circulating levels of melatonin are tightly linked to circadian rhythm, with melatonin levels lower during the day and higher during the night [68]. Prepubertal children produce higher levels of melatonin compared to post-pubertal peers, with both sexes showing a similar decline in melatonin levels across puberty; overall, melatonin levels tend to be higher in females during puberty [15]. Melatonin appears to have a role in keeping the HPG axis in a quiescent state during the peri-pubertal period, although the mechanism behind its role in the reawakening of the HPG axis is poorly understood [68].

The role of melatonin in the aetiopathogenesis of AIS is poorly understood but believed to be of neuroendocrine origin. Melatonin was initially implicated in AIS following pinealectomy of chickens and later replicated in bipedal rodents; however, these findings were not replicated in non-human primates, suggesting that the melatonin function varies between in primates and other mammals (reviewed in [21]). Melatonin levels in AIS patients are relatively inconclusive, with several studies reporting no differences between scoliotic patients and controls, making it difficult to draw any sound conclusions [20, 21]. A likely reason for melatonin acting as a cause of scoliosis in the chicken, but perhaps not in other species, is the distribution of the melatonin receptor; its distribution in chick is wider than in any other animal, with expression observed in dorsal grey matter of the spinal cord, the brainstem and the gonads [20].

Melatonin functions through binding to membrane-associated G-protein-coupled receptors, melatonin 1A (MTNR1B/MT1) and melatonin 1B (MTNR1B/MT2) [51]. Osteoblasts from AIS patients display dysfunctional melatonin signalling and reduced cell proliferation in response to melatonin treatment [6, 40, 41, 50]. However, while genetic variants in the *MTNR1B* gene promoter was previously been associated with AIS but not curve severity [59], this association not been replicated in additional genetic studies [71, 84].

Insulin-Like Growth Factor and Growth Hormone

Growth hormone (GH), responsible for longitudinal growth in children and adolescents, also has important roles in metabolism and protein synthesis. Increased sex steroid production during puberty stimulates GH and insulin-like growth factor (IGF-1) production, both required for bone and muscle growth [31, 47]. GH produced by the pituitary gland stimulates IGF-1 production by the liver (Fig. 8.1). During adolescence, GH and IGF-1 levels are higher compared to the prepubertal and adult life stages [38].

Irregular growth during pubertal growth spurt may be involved in progression and severity of AIS, prompting investigation into the role of the GH/IGF-1 axis in AIS. Previously, it was found that an *IGF-1* gene polymorphism (rs5742612) is associated with curve severity but not occurrence in Chinese [87] and Korean groups [49]. However, this SNP association has not been replicated in further studies, including an additional Han Chinese group [54, 71, 85]. Genetic association studies for the growth hormone receptor gene (*GHR*) also report similar findings, with no association found between *GHR* gene polymorphisms and either AIS onset or progression [60, 85]. Together these studies suggest that there is currently insufficient evidence to support a direct role of the GH/IGF-1 axis in AIS aetiology.

Metabolic Regulation of Puberty and AIS

Several studies found that the body mass index (BMI) of both girls and boys with AIS is lower compared to matched healthy peers [55, 67, 79]. Clark et al. [14] investigated the association between body weight, BMI and AIS in girls included as part of a longitudinal study, at two ages, 10 years (before scoliosis onset) and 15 years (5.9% now had scoliosis) [14]. Their findings not only revealed that a low BMI/body weight is associated with an increased risk of scoliosis, but those girls that later developed a scoliosis already had a lower BMI at 10 years of age [14]. Together these studies suggest that prepubertal body weight may be a contributing factor to scoliosis risk.

Puberty onset also requires the acquisition of a specific metabolic state and the presence of sufficient energy reserves. Both menarche and the adolescent growth spurt are delayed in girls with a low body weight [7, 27]. A key hormone required to signal sufficient energy reserves for pubertal development is leptin, a hormone also linked to AIS and low body weight.

Leptin Signalling

Leptin, an adipocyte-secreted signalling hormone, is responsible for maintaining bone health and energy metabolism [75]. Leptin levels increase during the prepubertal period, gradually rising in females during puberty; in contrast, males experience a transient spike in leptin, decreasing following the end of puberty [5, 9].

Several clinical studies have implicated leptin bioavailability in AIS aetiology. Circulating leptin levels are lower in individuals with scoliosis, compared to controls [34, 61]. Adjusting for age and menstrual status, AIS girls were typically taller (corrected height), had a lower body weight and BMI and exhibited a marked decrease circulating leptin levels [61, 72]. Given that the leptin levels in an individual is related to the proportion of adipose tissue [39], it is difficult to determine if the reduced leptin levels in AIS girls is a consequence reduced body fat and/or indicates a problem with the leptin pathway. In one cohort studied by Nikolova et al. [52], there was no significant association between leptin functional polymorphisms (rs7799039) and a susceptibility to idiopathic scoliosis, curve severity or pattern [52].

The *leptin receptor* gene (*LEPR*) encodes for several isoforms of the leptin receptor, which can be found at the plasma membrane of a large variety of cells [44]. It has a role in mediating leptin action by controlling the clearance of leptin and binding leptin in the bloodstream and aiding leptin transport across the blood brain barrier [8]. Decreased leptin receptor numbers leading to leptin hyposensitivity may provide an explanation for the lower leptin levels in some AIS patients [14, 32]. Tam et al. [72] observed higher serum LepR levels and lower free leptin index in AIS girls suggesting that altered leptin bioavailability could contribute to the observed lower body weight, lower BMI and abnormal body composition that appear to develop in scoliosis simultaneously [72]. Wang et al. [78] performed a case-control study examining the relative levels of serum leptin and leptin receptors in primary chondrocytes collected from scoliotic patients undergoing surgery [78]. AIS patient samples showed reduced membrane expression of the leptin receptor compared to control samples; however, this was not due to lower *LEPR* (*Ob-R*) mRNA expression [78]. Using in vitro assays, the authors found that LepR (also called Ob-R) endocytosis and membrane insertion may be the cause of lower plasma membrane LepR levels in AIS patient-derived chondrocytes.

The association of six *LEPR* gene polymorphisms and AIS in a Han Chinese cohort identified a single polymorphism, rs2767485, as being significantly associated with susceptibility to AIS but not curve severity [35]. This SNP has been previously identified as determinant of plasma LepR levels [70], although the mechanism lowering LepR levels is unknown; this is in accordance with the in vitro cell culture work by Wang and colleagues [78], further supporting that LepR metabolism is altered in some AIS patients.

Summary

Adolescence is a major period of sex maturation and growth for the human body and requires the successful integration of many hormonal pathways. Therefore, it is unsurprising that multiple pathways, whose activity dramatically changes during puberty and often in a sex-dimorphic fashion, have also been implicated in AIS. Although we have long suspected that AIS may be a multifactorial disease, research has largely focused on each aspect of AIS biology separately with some improvements to our understanding of the disease. The identification of a biological marker present during adolescence would permit early prognosis for individuals with AIS and progression risk, ultimately allowing for earlier stage intervention, avoiding invasive treatment at later stages of disease development. However, given the complexity of the underlying biology, particularly active during puberty, a single marker is unlikely to be sufficient. Recent studies are beginning to examine the compounding effects of possible genetic variants in AIS progression and disease, integrating this data with animal models and in vitro cell mechanistic studies will be important to unravel AIS aetiology.

References

1. Abreu AP, Kaiser UB. Pubertal development and regulation. Lancet Diabetes Endocrinol. 2016;4:254–64.
2. Adobor RD, Riise RB, Sorensen R, Kibsgard TJ, Steen H, Brox JI. Scoliosis detection, patient characteristics, referral patterns and treatment in the absence of a screening program in Norway. Scoliosis. 2012;7:18.
3. Aksglaede L, Olsen LW, Sorensen TI, Juul A. Forty years trends in timing of pubertal growth spurt in 157,000 Danish school children. PLoS One. 2008;3:e2728.
4. Arfai K, Pitukcheewanont PD, Goran MI, Tavare CJ, Heller L, Gilsanz V. Bone, muscle, and fat: sex-related differences in prepubertal children. Radiology. 2002;224:338–44.

5. Arslanian S, Suprasongsin C, Kalhan SC, Drash AL, Brna R, Janosky JE. Plasma leptin in children: relationship to puberty, gender, body composition, insulin sensitivity, and energy expenditure. Metabolism. 1998;47:309–12.
6. Azeddine B, Letellier K, Wang da S, Moldovan F, Moreau A. Molecular determinants of melatonin signaling dysfunction in adolescent idiopathic scoliosis. Clin Orthop Relat Res. 2007;462:45–52.
7. Baker ER. Body weight and the initiation of puberty. Clin Obstet Gynecol. 1985;28:573–9.
8. Banks WA. Leptin transport across the blood-brain barrier: implications for the cause and treatment of obesity. Curr Pharm Des. 2001;7:125–33.
9. Blum WF, Englaro P, Hanitsch S, Juul A, Hertel NT, Muller J, et al. Plasma leptin levels in healthy children and adolescents: dependence on body mass index, body fat mass, gender, pubertal stage, and testosterone. J Clin Endocrinol Metab. 1997;82:2904–10.
10. Canavese F, Dimeglio A. Normal and abnormal spine and thoracic cage development. World J Orthod. 2013;4:167–74.
11. Canavese F, Kaelin A. Adolescent idiopathic scoliosis: indications and efficacy of nonoperative treatment. Indian J Orthop. 2011;45:7–14.
12. Charles YP, Daures JP, de Rosa V, Dimeglio A. Progression risk of idiopathic juvenile scoliosis during pubertal growth. Spine. 2006;31:1933–42.
13. Choudhry MN, Ahmad Z, Verma R. Adolescent idiopathic scoliosis. Open Orthop J. 2016;10:143–54.
14. Clark EM, Taylor HJ, Harding I, Hutchinson J, Nelson I, Deanfield JE, et al. Association between components of body composition and scoliosis: a prospective cohort study reporting differences identifiable before the onset of scoliosis. J Bone Miner Res. 2014;29:1729–36.
15. Crowley SJ, Acebo C, Carskadon MA. Human puberty: salivary melatonin profiles in constant conditions. Dev Psychobiol. 2012;54:468–73.
16. Cui J, Shen Y, Li R. Estrogen synthesis and signaling pathways during aging: from periphery to brain. Trends Mol Med. 2013;19:197–209.
17. Demirkiran G, Dede O, Yalcin N, Akel I, Marcucio R, Acaroglu E. Selective estrogen receptor modulation prevents scoliotic curve progression: radiologic and histomorphometric study on a bipedal C57Bl6 mice model. Eur Spine J. 2014;23:455–62.
18. Dimeglio A, Canavese F. The growing spine: how spinal deformities influence normal spine and thoracic cage growth. Eur Spine J. 2012;21:64–70.
19. Dimeglio A, Canavese F, Bonnel F. Normal growth of the spine and thorax. Berlin/Heidelberg: Springer-Verlag; 2016.
20. Fagan AB, Kennaway DJ, Sutherland AD. Total 24-hour melatonin secretion in adolescent idiopathic scoliosis. A case-control study. Spine. 1998;23:41–6.
21. Girardo M, Bettini N, Dema E, Cervellati S. The role of melatonin in the pathogenesis of adolescent idiopathic scoliosis (AIS). Eur Spine J. 2011;20(Suppl 1):S68–74.
22. Granados A, Gebremariam A, Lee JM. Relationship between timing of peak height velocity and pubertal staging in boys and girls. J Clin Res Pediatr Endocrinol. 2015;7:235–7.
23. Herbison AE. Control of puberty onset and fertility by gonadotropin-releasing hormone neurons. Nat Rev Endocrinol. 2016;12:452–66.
24. Inoue M, Minami S, Nakata Y, Kitahara H, Otsuka Y, Isobe K, et al. Association between estrogen receptor gene polymorphisms and curve severity of idiopathic scoliosis. Spine. 2002;27:2357–62.
25. Iwamuro S, Sakakibara M, Terao M, Ozawa A, Kurobe C, Shigeura T, et al. Teratogenic and anti-metamorphic effects of bisphenol A on embryonic and larval Xenopus laevis. Gen Comp Endocrinol. 2003;133:189–98.
26. Janusz P, Kotwicka M, Andrusiewicz M, Czaprowski D, Czubak J, Kotwicki T. Estrogen receptors genes polymorphisms and age at menarche in idiopathic scoliosis. BMC Musculoskelet Disord. 2014;15:383.
27. Kaplowitz PB. Link between body fat and the timing of puberty. Pediatrics. 2008;121(Suppl 3):S208–17.
28. Konieczny MR, Senyurt H, Krauspe R. Epidemiology of adolescent idiopathic scoliosis. J Child Orthop. 2013;7:3–9.

29. Kouwenhoven JW, Castelein RM. The pathogenesis of adolescent idiopathic scoliosis: review of the literature. Spine. 2008;33:2898–908.
30. Leboeuf D, Letellier K, Alos N, Edery P, Moldovan F. Do estrogens impact adolescent idiopathic scoliosis? Trends Endocrinol Metab. 2009;20:147–52.
31. Leung KC, Johannsson G, Leong GM, Ho KK. Estrogen regulation of growth hormone action. Endocr Rev. 2004;25:693–721.
32. Liang G, Gao W, Liang A, Ye W, Peng Y, Zhang L, Sharma S, Su P, Huang D. Normal leptin expression, lower adipogenic ability, decreased leptin receptor and hyposensitivity to leptin in adolescent idiopathic scoliosis. PLoS One. 2012;7:e36648.
33. Little DG, Song KM, Katz D, Herring JA. Relationship of peak height velocity to other maturity indicators in idiopathic scoliosis in girls. J Bone Joint Surg Am. 2000;82:685–93.
34. Liu Z, Tam EM, Sun GQ, Lam TP, Zhu ZZ, Sun X, et al. Abnormal leptin bioavailability in adolescent idiopathic scoliosis: an important new finding. Spine. 2012;37:599–604.
35. Liu Z, Wang F, Xu LL, Sha SF, Zhang W, Qiao J, et al. Polymorphism of rs2767485 in leptin receptor gene is associated with the occurrence of adolescent idiopathic scoliosis. Spine. 2015;40:1593–8.
36. Loncar-Dusek M, Pecina M, Prebeg Z. A longitudinal study of growth velocity and development of secondary gender characteristics versus onset of idiopathic scoliosis. Clin Orthop Relat Res. 1991:278–82.
37. Lonstein JE, Carlson JM. The prediction of curve progression in untreated idiopathic scoliosis during growth. J Bone Joint Surg Am. 1984;66:1061–71.
38. Luna AM, Wilson DM, Wibbelsman CJ, Brown RC, Nagashima RJ, Hintz RL, et al. Somatomedins in adolescence: a cross-sectional study of the effect of puberty on plasma insulin-like growth factor I and II levels. J Clin Endocrinol Metab. 1983;57:268–71.
39. Maffei M, Halaas J, Ravussin E, Pratley RE, Lee GH, Zhang Y, et al. Leptin levels in human and rodent: measurement of plasma leptin and ob RNA in obese and weight-reduced subjects. Nat Med. 1995;1:1155–61.
40. Man GC, Wang WW, Yeung BH, Lee SK, Ng BK, Hung WY, et al. Abnormal proliferation and differentiation of osteoblasts from girls with adolescent idiopathic scoliosis to melatonin. J Pineal Res. 2010;49:69–77.
41. Man GC, Wong JH, Wang WW, Sun GQ, Yeung BH, Ng TB, et al. Abnormal melatonin receptor 1B expression in osteoblasts from girls with adolescent idiopathic scoliosis. J Pineal Res. 2011;50:395–402.
42. Mannion AF, Meier M, Grob D, Muntener M. Paraspinal muscle fibre type alterations associated with scoliosis: an old problem revisited with new evidence. Eur Spine J. 1998;7:289–93.
43. Mao SH, Jiang J, Sun X, Zhao Q, Qian BP, Liu Z, et al. Timing of menarche in Chinese girls with and without adolescent idiopathic scoliosis: current results and review of the literature. Eur Spine J. 2011;20:260–5.
44. Margetic S, Gazzola C, Pegg GG, Hill RA. Leptin: a review of its peripheral actions and interactions. Int J Obes Relat Metab Disord. 2002;26:1407–33.
45. Maria S, Witt-Enderby PA. Melatonin effects on bone: potential use for the prevention and treatment for osteopenia, osteoporosis, and periodontal disease and for use in bone-grafting procedures. J Pineal Res. 2014;56:115–25.
46. Marshall WA, Tanner JM. Growth and physiological development during adolescence. Annu Rev Med. 1968;19:283–300.
47. Mauras N, Rogol AD, Haymond MW, Veldhuis JD. Sex steroids, growth hormone, insulin-like growth factor-1: neuroendocrine and metabolic regulation in puberty. Horm Res. 1996;45:74–80.
48. Mendis-Handagama SM, Ariyaratne HB. Differentiation of the adult Leydig cell population in the postnatal testis. Biol Reprod. 2001;65:660–71.
49. Moon ES, Kim HS, Sharma V, Park JO, Lee HM, Moon SH, et al. Analysis of single nucleotide polymorphism in adolescent idiopathic scoliosis in Korea: for personalized treatment. Yonsei Med J. 2013;54:500–9.
50. Moreau A, Wang DS, Forget S, Azeddine B, Angeloni D, Fraschini F, et al. Melatonin signaling dysfunction in adolescent idiopathic scoliosis. Spine. 2004;29:1772–81.

51. Ng KY, Leong MK, Liang H, Paxinos G. Melatonin receptors: distribution in mammalian brain and their respective putative functions. Brain Struct Funct. 2017;222:2921.
52. Nikolova S, Yablanski V, Vlaev E, Getova G, Atanasov V, Stokov L, et al. In search of biomarkers for idiopathic scoliosis: leptin and BMP4 functional polymorphisms. J Biomark. 2015a;2015:425310.
53. Nikolova S, Yablanski V, Vlaev E, Stokov L, Savov A, Kremensky I. Association between estrogen receptor alpha gene polymorphisms and susceptibility to idiopathic scoliosis in Bulgarian patients: a case-control study. Open Access Maced J Med Sci. 2015b;3:278–82.
54. Nikolova S, Yablanski V, Vlaev E, Stokov L, Savov AS, Kremensky IM. Association study between idiopathic scoliosis and polymorphic variants of VDR, IGF-1, and AMPD1 genes. Genet Res Int. 2015c;2015:852196.
55. Normelli H, Sevastik J, Ljung G, Aaro S, Jonsson-Soderstrom AM. Anthropometric data relating to normal and scoliotic Scandinavian girls. Spine. 1985;10:123–6.
56. Onaolapo OJ, Onaolapo AY. Melatonin, adolescence, and the brain: an insight into the period-specific influences of a multifunctional signaling molecule. Birth Defects Res. 2017;109:1659–71.
57. Peng Y, Liang G, Pei Y, Ye W, Liang A, Su P. Genomic polymorphisms of G-protein estrogen receptor 1 are associated with severity of adolescent idiopathic scoliosis. Int Orthop. 2012;36:671–7.
58. Prossnitz ER, Barton M. The G-protein-coupled estrogen receptor GPER in health and disease. Nat Rev Endocrinol. 2011;7:715–26.
59. Qiu XS, Tang NL, Yeung HY, Lee KM, Hung VW, Ng BK, et al. Melatonin receptor 1B (MTNR1B) gene polymorphism is associated with the occurrence of adolescent idiopathic scoliosis. Spine. 2007a;32:1748–53.
60. Qiu XS, Tang NL, Yeung HY, Qiu Y, Cheng JC. Genetic association study of growth hormone receptor and idiopathic scoliosis. Clin Orthop Relat Res. 2007b;462:53–8.
61. Qiu Y, Sun X, Qiu X, Li W, Zhu Z, Zhu F, et al. Decreased circulating leptin level and its association with body and bone mass in girls with adolescent idiopathic scoliosis. Spine. 2007c;32:2703–10.
62. Saggese G, Baroncelli GI, Bertelloni S. Puberty and bone development. Best Pract Res Clin Endocrinol Metab. 2002;16:53–64.
63. Sharma G, Prossnitz ER. GPER/GPR30 knockout mice: effects of GPER on metabolism. Methods Mol Biol. 2016;1366:489–502.
64. Sharma G, Prossnitz ER. G-Protein-Coupled Estrogen Receptor (GPER) and sex-specific metabolic homeostasis. Adv Exp Med Biol. 2017;1043:427–53.
65. Shi B, Mao S, Liu Z, Sun X, Zhu Z, Zhu F, et al. Spinal growth velocity versus height velocity in predicting curve progression in peri-pubertal girls with idiopathic scoliosis. BMC Musculoskelet Disord. 2016;17:368.
66. Simpson ER. Aromatization of androgens in women: current concepts and findings. Fertil Steril. 2002;77(Suppl 4):S6–10.
67. Siu King Cheung C, Tak Keung Lee W, Kit Tse Y, Ping Tang S, Man Lee K, et al. Abnormal peri-pubertal anthropometric measurements and growth pattern in adolescent idiopathic scoliosis: a study of 598 patients. Spine. 2003;28:2152–7.
68. Srinivasan V, Spence WD, Pandi-Perumal SR, Zakharia R, Bhatnagar KP, Brzezinski A. Melatonin and human reproduction: shedding light on the darkness hormone. Gynecol Endocrinol. 2009;25:779–85.
69. Stokes IA, Windisch L. Vertebral height growth predominates over intervertebral disc height growth in adolescents with scoliosis. Spine. 2006;31:1600–4.
70. Sun Q, Cornelis MC, Kraft P, Qi L, van Dam RM, Girman CJ, et al. Genome-wide association study identifies polymorphisms in LEPR as determinants of plasma soluble leptin receptor levels. Hum Mol Genet. 2010;19:1846–55.
71. Takahashi Y, Matsumoto M, Karasugi T, Watanabe K, Chiba K, Kawakami N, et al. Lack of association between adolescent idiopathic scoliosis and previously reported single nucleotide

polymorphisms in MATN1, MTNR1B, TPH1, and IGF1 in a Japanese population. J Orthop Res. 2011;29:1055–8.
72. Tam EM, Liu Z, Lam TP, Ting T, Cheung G, Ng BK, et al. Lower muscle mass and body fat in adolescent idiopathic scoliosis are associated with abnormal leptin bioavailability. Spine. 2016;41:940–6.
73. Tang NL, Yeung HY, Lee KM, Hung VW, Cheung CS, Ng BK, et al. A relook into the association of the estrogen receptor [alpha] gene (PvuII, XbaI) and adolescent idiopathic scoliosis: a study of 540 Chinese cases. Spine. 2006;31:2463–8.
74. Tanner JM, Whitehouse RH, Marubini E, Resele LF. The adolescent growth spurt of boys and girls of the Harpenden growth study. Ann Hum Biol. 1976;3:109–26.
75. Thomas T, Gori F, Khosla S, Jensen MD, Burguera B, Riggs BL. Leptin acts on human marrow stromal cells to enhance differentiation to osteoblasts and to inhibit differentiation to adipocytes. Endocrinol. 1999;140:1630–8.
76. Ueno M, Takaso M, Nakazawa T, Imura T, Saito W, Shintani R, et al. A 5-year epidemiological study on the prevalence rate of idiopathic scoliosis in Tokyo: school screening of more than 250,000 children. J Orthop Sci. 2011;16:1–6.
77. Wang W, Wang Z, Zhu Z, Zhu F, Qiu Y. Body composition in males with adolescent idiopathic scoliosis: a case-control study with dual-energy X-ray absorptiometry. BMC Musculoskelet Disord. 2016a;17:107.
78. Wang YJ, Yu HG, Zhou ZH, Guo Q, Wang LJ, Zhang HQ. Leptin receptor metabolism disorder in primary chondrocytes from adolescent idiopathic scoliosis girls. Int J Mol Sci. 2016b;17. pii: E1160.
79. Wei-Jun W, Xu S, Zhi-Wei W, Xu-Sheng Q, Zhen L, Yong Q. Abnormal anthropometric measurements and growth pattern in male adolescent idiopathic scoliosis. Eur Spine J. 2012;21:77–83.
80. Weinstein SL, Ponseti IV. Curve progression in idiopathic scoliosis. J Bone Joint Surg Am. 1983;65:447–55.
81. Weiss HR, Moramarco MM, Borysov M, Ng SY, Lee SG, Nan X, et al. Postural rehabilitation for adolescent idiopathic scoliosis during growth. Asian Spine J. 2016;10:570–81.
82. Wynne-Davies R. Familial (idiopathic) scoliosis. A family survey. J Bone Joint Surg Br. 1968;50:24–30.
83. Yang M, Li C, Li M. The estrogen receptor alpha gene (XbaI, PvuII) polymorphisms and susceptibility to idiopathic scoliosis: a meta-analysis. J Orthop Sci. 2014;19:713–21.
84. Yang M, Wei X, Yang W, Li Y, Ni H, Zhao Y, et al. The polymorphisms of melatonin receptor 1B gene (MTNR1B) (rs4753426 and rs10830963) and susceptibility to adolescent idiopathic scoliosis: a meta-analysis. J Orthop Sci. 2015;20:593–600.
85. Yang Y, Wu Z, Zhao T, Wang H, Zhao D, Zhang J, et al. Adolescent idiopathic scoliosis and the single-nucleotide polymorphism of the growth hormone receptor and IGF-1 genes. Orthopedics. 2009;32:411.
86. Yasar P, Ayaz G, User SD, Gupur G, Muyan M. Molecular mechanism of estrogen-estrogen receptor signaling. Reprod Med Bio. 2017;16:4–20.
87. Yeung HY, Tang NL, Lee KM, Ng BK, Hung VW, Kwok R, et al. Genetic association study of insulin-like growth factor-I (IGF-I) gene with curve severity and osteopenia in adolescent idiopathic scoliosis. Stud Health Technol Inform. 2006;123:18–24.
88. Ylikoski M. Growth and progression of adolescent idiopathic scoliosis in girls. J Pediatr Orthop B. 2005;14:320–4.
89. Zamecnik J, Krskova L, Hacek J, Stetkarova I, Krbec M. Etiopathogenesis of adolescent idiopathic scoliosis: expression of melatonin receptors 1A/1B, calmodulin and estrogen receptor 2 in deep paravertebral muscles revisited. Mol Med Rep. 2016;14:5719–24.
90. Zhao L, Roffey DM, Chen S. Association between the Estrogen Receptor Beta (ESR2) Rs1256120 single nucleotide polymorphism and adolescent idiopathic scoliosis: a systematic review and meta-analysis. Spine. 2017;42:871–8.

Index

K. Kusumi, S. L. Dunwoodie (eds.), *The Genetics and Development of Scoliosis*, https://doi.org/10.1007/978-3-319-90149-7

Printed in the United States
By Bookmasters